材料科学与工程进展

上海材料前沿论坛2020年春季会议大摘要

Progress in Materials Science and Engineering

Extended Abstract for Shanghai Forum
on Materials Frontier 2020 Spring

董 瀚 朱 波 主编

上海大学出版社
·上海·

图书在版编目(CIP)数据

材料科学与工程进展：上海材料前沿论坛2020年春季会议大摘要/董瀚，朱波主编. —上海：上海大学出版社，2021.1（2021.11重印）
 ISBN 978-7-5671-4150-6

Ⅰ.①材… Ⅱ.①董…②朱… Ⅲ.①材料科学-文集 Ⅳ.①TB3-53

中国版本图书馆 CIP 数据核字(2021)第 001514 号

责任编辑 李　双
封面设计 柯国富
技术编辑 金　鑫　钱宇坤

材料科学与工程进展

上海材料前沿论坛2020年春季会议大摘要
Progress in Materials Science and Engineering
Extended Abstract for Shanghai Forum
on Materials Frontier 2020 Spring
董　瀚　朱　波　主编
上海大学出版社出版发行
（上海市上大路99号　邮政编码 200444）
(http://www.shupress.cn 发行热线 021-66135112)
出版人　戴骏豪
*
南京展望文化发展有限公司排版
上海华业装璜印刷厂有限公司印刷　各地新华书店经销
开本 787mm×1092mm　1/16　印张 18.5　字数 405 千
2021年1月第1版　2021年11月第2次印刷
ISBN 978-7-5671-4150-6/TB·20　定价　68.00元

版权所有　侵权必究
如发现本书有印装质量问题请与印刷厂质量科联系
联系电话：021-56475919

材料科学与工程进展

上海材料前沿论坛 2020 年春季会议

Progress in Materials Science and Engineering

Shanghai Forum on Materials Frontier 2020 Spring

上海大学

Shanghai University

2020 年 4 月 28 日，线上会议

April 28, 2020, Online

一、会议主题

聚焦材料科学与工程领域的发展前沿,研讨学术界与工业界科研热点与关键技术问题。

1. 先进金属材料及工程

包括但不局限于钢铁材料和有色金属材料的设计、生产、零件加工、服役评价等。

2. 冶金工程技术

包括但不局限于钢铁冶金、智能冶金、电磁冶金、有色冶金及资源利用技术等。

3. 功能材料及技术

包括但不局限于智能无机材料、薄膜半导体材料与器件、光电材料与器件、热电材料与器件、先进能源材料、纳米功能材料、纳米结构与催化、环境材料等。

4. 软物质材料

包括但不局限于仿生高分子材料、生物医用材料、有机生物电子材料、智能凝胶、超分子组装、柔性功能材料与器件、高分子化学、高分子化工、复合材料等。

5. 纤维与低维材料

包括但不局限于高性能纤维与复合材料、功能纤维、光导纤维、超弹性导电纤维、有机-无机杂化低维材料、能量存储与转换材料、共轭高分子、有机多孔材料、二维高分子等。

6. 材料基因工程

包括但不局限于高通量实验、高通量计算、数据挖掘、材料大数据及数据库、材料信息学等。

7. 长三角材料技术产业链

新型耐蚀钢及配套材料技术推广应用、紧固件及材料技术产业链的建设完善、市场分析及预测、团体标准应用情况及质量控制。

二、主办单位

上海大学
中国工程科技知识中心材料分中心

三、协办单位

中国内燃机学会新材料与表面技术分会
东华大学 纤维材料改性国家重点实验室
上海工程技术大学
江苏冶金技术研究院

上海宝钢不锈钢有限公司、宝钢特钢有限公司
江苏集萃安泰创明先进能源材料研究院有限公司
上大新材料(泰州)研究院有限公司
腾讯云计算(北京)有限责任公司

四、会议组织机构

会议名誉主席： 干勇

会议主席： 刘昌胜

学术委员会： David A. Weitz、丁文江、干勇、叶明新、朱向阳、朱美芳、刘玠、刘昌胜、刘浏、江莞、孙宝德、孙晋良、李劲、李春忠、吴明红、张弛、张统一、张荻、林绍梁、周邦新、周国治、周细应、单爱党、俞建勇、徐滨士、韩伟、鲁雄刚(按姓氏笔画排列)

组织委员会： 马敬红、王刚、王江、王林军、王勇、韦习成、尹静波、吕战鹏、朱波、任忠鸣、刘文庆、刘福窑、许茜、杜娟、李文献、李红、李谋成、李喜、李谦、杨健、吴晓春、张阿方、张金仓、张登松、张耀鹏、钟云波、施利毅、施思齐、施鹰、骆军、高彦峰、黄健、董瀚、鲁晓刚、廖耀祖、翟启杰、颜世峰(按姓氏笔画排列)

会议秘书处

秘书长：董瀚、朱波

副秘书长：廖耀祖、卫慧

学术领域秘书：赵洪山、彭伟、廉心桐、王春燕、李传军、邹星礼、帅三三、张银辉、徐龙云、吴思炜、郭凯、陈宏飞、崔苑苑、李亚捷、贾林、张坤玺、耿志、吴青、黄波、赖建明、姜沪、陆恒昌、於乾英、滕晓黎、吴红敏、罗娟娟

网络会议系统支持： 杭坚[腾讯云计算(北京)有限责任公司]

序

材料是人类社会发展的物质基础。一代材料,造就一代装备,引领了一代产业发展。随着人们认识世界的不断深入和需求的持续提高,材料不断发展演变,并构成了今天的变化万千体系。今天的材料就像一棵枝繁叶茂的大树,品种繁多,数不胜数。针对社会发展需求,逐渐形成了先进基础材料、关键战略材料、前沿新材料等三大体系材料,受到了人们的普遍关注。新材料是我国的战略性新兴产业之一,支撑了经济与社会发展。

由于突如其来的新冠疫情,2020 年春天人们被禁足在家,然而却禁不住材料人相互交流的愿望。2020 年 4 月 28 日,上海材料前沿论坛 2020 年春季会议(SHMat 2020 Spring)暨上海高校国际青年学者论坛首次线上"云 TALK"模式召开。大家一起来到了天空云上连线聚会,探讨了材料科学、技术与工程的最新进展,同时也在探索新的学术交流方式。

论坛非常荣幸地邀请到了中国科学院院士陈学思、刘昌胜、张统一、朱美芳,中国工程院院士刘玠,1987 年诺贝尔化学奖得主 Jean-Marie Lehn 教授以及日本工程院院士庄子哲雄等来自中国、英国、法国、瑞士、丹麦、日本、澳大利亚、美国等国家的百余位行内知名专家和国际青年学者。他们齐聚一堂,探讨了材料领域的科技进展。论坛共有 144 场学术报告(78 场邀请报告和 66 场青年学者报告)以及 25 份 Poster 展示,其中 32 位报告人和 4 位 Poster 展示者来自海外。近 800 人次进入云上会议室,2 200 多人同时在线观看 1 个主论坛和 7 个分会会议视频直播。

本次云论坛由上海大学和中国工程科技知识中心材料分中心主办,上海市教委指导与支持。中国工程院院士干勇担任大会名誉主席,中国科学院院士、上海大学校长刘昌胜担任大会主席,上海大学党委书记成旦红向大会致辞,上海大学材料科学与工程学院院长董瀚主持开幕式。

先进金属材料及工程分会探讨了强度极限化、多功能化、材料制备、服役失效评价等主题。冶金工程技术分会探讨了金属冶金、电磁冶金、智能冶金、资源利用等主题。功能材料及技术分会探讨了光电材料、能源电池材料、核辐射探测材料、催化材料、纳米复合材料、形状记忆合金等主题。软物质材料分会探讨了生命体系、高分子、液晶态、凝胶态、复合功能物质等主题。纤维与低维材料分会探讨了高性能纤维和复合材料、功能纤维材料、光纤维材料等主题。材料基因工程分会研讨了高通量实验、高通量计算、数据挖掘、材料大数据及数据库、材料信息学等主题。长三角材料技术产业链分会探讨了紧固件与材料、耐蚀钢生产与应用等技术产业链构建可能性。

感谢为计算机互联网信息技术做出贡献的科技人员，也要感谢我们的材料同行，为计算机互联网信息的发展奠定了材料基础，比如硅单晶芯片材料、光纤通信材料、高纯净度靶材与显示材料、机械加工材料等。当然更要感谢所有的材料人，有了他们才有了今天的丰富多彩的物质世界。材料发展到了今天就像一颗参天大树，枝繁叶茂，并且还在不断地生长出更多更好的新材料。

"上海材料前沿论坛"将立足于上海科创中心建设与长三角一体化发展国家战略提供的时代机遇，持续每年线上线下召开，聚焦前沿成果和趋势，聚集海内外科学家与工程师，搭建一流科技交流互动平台，促进材料科技发展。

上海大学校长、中国科学院院士

刘昌胜 教授

Preface

Materials form the basis of human society. It is believed that every new generation of materials may create one new generation of device, and then lead to a new generation of manufacture. Based on the demands for materials and the understandings on science, materials have been developed remarkably, spanned multiple disciplines and covered a broad range of applications. Noteworthily speaking, three categories of materials, including advanced essential materials, key strategic materials, and cutting-edge new materials, have attracted significant research attentions. New materials have been identified as one of the new strategic industries in China, which will support economics and social development.

Due to the lock down caused by COVID-19, Shanghai Forum on Materials Frontier (SHMat) 2020 Spring was courageously held online on April 28, 2020. Significantly, the forum exchanged the latest research advances in materials science, technology, and engineering in this special time, and also manifested a new mode of academic conference.

The SHMat 2020 Spring honorably invited hundreds of international scholars, including the Academicians of the Chinese Academy of Science (Prof. Chen Xuesi, Prof. Liu Changsheng, Prof. Zhang Tongyi, and Prof. Zhu Meifang), the Academician of the Chinese Academy of Engineering (Prof. Liu Jie), the 1987 Nobel Prize laureate in chemistry (Prof. Jean-Marie Lehn), the Academician of the Japanese Academy of Engineering (Prof. Tetsuo Shoji), and other renowned and young scholars from China, UK, France, Switzerland, Denmark, Japan, Australia, and USA, to join this forum online. Furthermore, 144 scientific reports (including 78 invited reports and 66 young scientist reports) and 25 posters were presented at the forum. Among them, 32 reporters and 4 poster presenters were from foreign countries. During the forum, almost 800 scholars joined the online meeting and at the same time 2,000 audiences watched the videos of the main forum live and seven branch meetings online.

The SHMat 2020 Spring was sponsored by Shanghai University and China Knowledge Centre for Engineering Sciences and Technology, Materials Branch; and it was guided by Shanghai Municipal Education Commission. The Honorary Chairperson of the Forum was Prof. Gan Yong (the Academician of the Chinese Academy of Engineering). The Forum Chairperson was Prof. Liu Changsheng (the Academician of the Chinese Academy of Science, the President of Shanghai University). Prof. Cheng Danhong (the Secretary of the CPC Committee of Shanghai University) delivered the opening speech. Prof. Dong Han (the Dean of School of Materials Science and Engineering) chaired the opening ceremony of this forum.

The SHMat 2020 Spring consisted of seven parallel sessions. The session of Advanced Metal Materials focused on the ultimate strength, multi-functionalization, manufacturing, and the evaluation of ferrous and non-ferrous metals, and etc. The session of Metallurgical Engineering included the metal metallurgy, electromagnetic metallurgy, artificial intelligence, and etc. The session of Functional Materials talked over the development of optoelectronic materials, energy materials, nanomaterials, catalysis materials, environmental materials, and etc. The session of Soft Matter Materials focused on biomimetic polymeric materials, liquid crystal materials, gel materials, composite materials, and etc. The session of Fibers and Low-dimensional Materials concentrated on high-performance fibers and composite materials, functional fibers, optical fibers, and etc. Moreover, the session of Materials Genome Initiatives focused on the high-throughput computing and experimental aspects, big data, materials informatics, and etc. The parallel session of Yangtze-Delta Materials Industry Chain focused on the production and application chain of fastener steel and corrosion-resistant steel.

The SHMat 2020 Spring greatly acknowledged the support from the technical community. Moreover, I am also grateful to all the materials scientists and engineers, who have actively participated in the Forum. The SHMat, promoted by the construction of Science and Innovation Center in Shanghai and the Yangtze River Delta, will be held every year. Undoubtedly, it will provide exchange opportunities on the cutting-edge achievements and development trends in materials fields, offer a platform to gather international outstanding scientists and engineers in materials fields, and establishe a world-class academic exchange and interaction platform to promote academic exchanges, idea collisions, and research cooperations.

Prof. *Liu Changsheng*
President of Shanghai University
Academician of the Chinese Academy of Science

目录 | Contents

第一部分　大会报告
Section 1　Plenary Lecture

新材料高质量发展战略研究 ··· 003
创新生物材料：支撑人类医疗方式变革 ································· 004
Development of SCC Resistant Austenitic Stainless Steels for LWR Application
　with a Special Emphasis on the Role of Hydrogen ················ 005
Dynamic Materials Towards Functional Adaptive Materials ·········· 009

第二部分　金属材料
Section 2　Metal Materials

体心立方金属中晶界结构的微观相场模拟研究 ······················· 013
Point Defects in Transparent Conducting Oxides ······················ 015
Antibacterial Mg Alloys Through Microalloying ······················ 017
Amorphous Al-based Metallic Plastics ································ 018
Investigating the Compatibility of Alumina-forming Alloys with Aggressive
　Pb Environments for Energy Related Application ················ 021
Fatigue Behavior and Its Influencing Factors of Ti-6Al-4V Fabricated by Additive
　Manufacturing：An Overview ·· 023
Temperature Capability Prediction of Nickel-Based Superalloys By Cluster-Plus-
　Glue-Atom Model ··· 025
The Effect of Fiber Laser Welding on Solute Segregation and Proprieties of
　CoCrCuFeNi High Entropy Alloy ···································· 028
The Corrosion Behaviour of Magnetocaloric Alloys La(Fe，Mn，Si)$_{13}$H$_x$ under
　Magnetic Field Conditions ··· 030
Phase Formation，Transformation and Stability in Micro-alloyed Sn-based Lead-
　free Solder Alloys and Joints ·· 032

基于先进核能技术需求的特种合金材料研制 ………………………………………… 033
Microstructure, Micro-chemistry and Stress Corrosion Cracking of Dissimilar
　　Metal Weld 16MND5/309L/308L/Z2CND18-12N Used in Nuclear Power
　　Plants ……………………………………………………………………………… 034
钢铁材料性能极限化探讨与初步结果：强度、塑性、韧性 …………………………… 037
Nb Microalloyed Ultrafine High C Spring Strip and Its 50 Year Evolution ……… 039
Microstructure Evolution and Crack Propagation of Pearlitic Wheel-rail Steel
　　under Service Condition ………………………………………………………… 042
双相钢损伤形核的介观起源 ……………………………………………………………… 044
The Size Effect of κ-carbides Precipitation on Mechanical and Hydrogen
　　Embrittlement Properties of a Fe-Mn-Al-C Low Density Steel ……………… 046
含 2% Mn 多相组织（M3）低合金钢中纳米 Cu 析出极限强化研究 ………………… 048
Study on the Applicability of Warm Forming Ultrahigh Strength Medium-Mn
　　Steel …………………………………………………………………………………… 051
Strengthening and Toughening Mechanism for High-alloyed Martensitic Steel …… 053
Physical Metallurgy-guided Machine Learning and Artificial Intelligent Design
　　of Ultrahigh-strength Stainless Steel …………………………………………… 057
Effect of Surface Deformation on Stress Corrosion Crack Initiation in
　　Austenitic Stainless Steels in PWR Primary Water …………………………… 059
刀具用高碳马氏体不锈钢碳化物调控研究 ……………………………………………… 061
High Nitrogen Steels: Livelihood Applications ………………………………………… 063
Investigations on the Stress Corrosion Cracking Initiation Behavior of 316LN
　　Stainless Steel in PWR Primary Water ………………………………………… 065
High Energy Storage of a Metallic Glass via Long-term Cryogenic Thermal
　　Cycling ……………………………………………………………………………… 069
Fabrication of Metal Matrix Composites by Solid-state Cold Spraying Process …… 072
One-step Sintering Synthesis of Superfine $L1_0$-FePt Nanoparticle by Using
　　Liquid-assisted …………………………………………………………………… 074
Metal-Organic Nanoprobe for Ratiometric Sensing Peroxynitrite and Hypochlorite
　　Through FRET …………………………………………………………………… 076

第三部分　冶金新技术
Section 3　Metallurgical Engineering

转炉炼钢底喷粉技术进展 ………………………………………………………………… 081
Current Situations and Future Perspective of Ironmaking Industry and Green
　　Ironmaking Technology ………………………………………………………… 083

Technology of Fine Bubbles Generated by Argon Injecting into the Down-snorkel of RH Degasser ………………………………………………………… 086

High-speed Mold Width Adjustment Technology and Application of Double C ……… 088

A Design of Self-generated Ti-Al-Si Gradient Coatings on Ti-6Al-4V Alloy Based on Silicon Concentration Gradient ……………………………………… 089

Nucleation and Growth Behavior of FePt Nanomaterials under High Magnetic Field ……………………………………………………………………… 092

Crystal Structure and Thermal Stability of the Intermetallic Phases in High-pressure Die Casting Mg-Al-RE Based Alloys ……………………………… 094

Study on the Microstructure and Mechanical Performances of C-X Steel Processed by Selective Laser Melting (SLM) Technology ………………………………… 096

金属凝固过程与均质化技术 …………………………………………………… 099

Modelling of the High Pressure Die Casting Process ……………………………… 101

Some Considerations for the New Generation of High-efficiency Continuous Casting Technology Development …………………………………………… 104

连铸坯中夹杂物成分空间分布的预报 ………………………………………… 107

机器学习模型预测钢铁冶金中的组织演变 …………………………………… 109

高温长时蠕变后G115钢微观组织的演变及其对硬度的影响 ………………… 111

The Formation Mechanism of Dislocation Patterns under Low Cycle Fatigue of High-manganese Austenitic TRIP Steel with Dominating Planar Slip Mode …… 114

Theory and Application of High-value Utilization of Typical Bulk Industrial Solid Waste ……………………………………………………………………… 115

第四部分 功能材料
Section 4 Functional Materials

Effect of Nanoparticles Al_2O_3 and Rare Earth on New Alumina Dispersion-strengthened Copper Alloy ………………………………………………… 121

Focusing Nano Metals Under Pressure ……………………………………………… 124

A General Mechanism of Grain Growth: Theory and Experimental …………… 126

Transition Metal Oxides Induced Spin-Orbit Coupling at Surface Conducting Diamond …………………………………………………………………… 128

The Preparation, Characterization and Application of Nanostructured Mg-based Hydrogen Storage Materials ……………………………………… 132

Design of Sulfur Cathode Host Material Based on Strong Polarity …………… 134

A Facile Applicable Strategy for Construction of 3D Porous Gelatin-Alginate Hydrogels for Deep Second-Degree Scald Wound Healing ………………… 136

Microstructures and Optoelectronic Properties of NiO Films Deposited by High Power Impulse Magnetron Sputtering ······ 138

Pressure-induced Dramatic Changes in Halide Perovskites ······ 140

Topological Transformation of Layered Double Hydroxide Nanosheets for Efficient Photocatalytic CO_2 reduction ······ 141

High-efficiency CO_2 Separation via Reaction-promoted Transport in Two-dimensional Sub-nanometer Channels ······ 144

锌离子电池与储能变色双功能器件 ······ 147

Explore Potential Advantages of Na-Ion Batteries with Ultralow-Concentration Electrolyte ······ 149

Study of TiO_2 Coated α-Fe_2O_3 Composites and the Oxygen-defects Effect on the Application as Anode of High-performance Li-ion Battery ······ 151

Design of Advanced Porous Materials for Sodium-ion Batteries ······ 153

纤维电子器件的连续化制备与实用性评价研究 ······ 156

Field-effect Control of Emergent Properties in Low Dimensional Quantum Materials ······ 158

Stretchable Conductive Nonwoven Fabrics with Self-cleaning Capability for Tunable Wearable Strain Sensor ······ 159

Novel Solid-state Elastocaloric Materials for Eco-refrigeration ······ 161

Design and Development of Advanced Transparent Insulation Materials for Energy Efficient Windows ······ 163

Smart Nanothermochromic Window Advances and Implications for Use in Different Climates ······ 166

Ultralight Programmable Bioinspired Aerogels with Integrated Multifunctionalities via Co-assembly ······ 168

第五部分　软物质材料
Section 5　Soft Mater Materials

Unlocking Polymer Degradability Through Mechanophore Activation ······ 173

Design of Polymers for Biomedical Applications ······ 176

Polypeptide Superhelices: Chirality and Morphology ······ 178

Renewable and Degradable Polymeric Materials Based on Coordination Ring-opening Polymerization ······ 180

Polymer Brush Based Inorganic-Organic Hybrid Materials ······ 181

Tetrapod Polymersomes ······ 184

Polymers/Carbon Nanomaterial Composites and Electrochemical Energy Storage ······ 186

Energy Storage and Conversion Using Conjugated Microporous Polymers ……… 187
聚合物分子刷及其功能材料 …………………………………………………… 188
嵌段共轭聚物的三维螺型结构 ………………………………………………… 189
Bio-inspired Mechano-functional Gels Through Multi-phase Order-structure
　　Engineering ……………………………………………………………………… 191
Polysaccharide-based Recoverable Double-network Hydrogel with High Strength
　　and Self-healing Properties ……………………………………………………… 192
A Self-healing Hydrogel with Pressure Sensitive Photoluminescence for Remote
　　Force Measurement and Healing Assessment ………………………………… 194
Preparation of Polyamino Acid Self-Healing Hydrogels Based on 2-Ureido-
　　4[1H]-Pyrimidinone …………………………………………………………… 198
Injectable Simvastatin-loaded Micelle/Hydrogel Composites for Bone Tissue
　　Engineering ……………………………………………………………………… 201
Symbiont Mutagenesis and Characterization as a Potential Method to Alter
　　Holobiont Stress Tolerance：A Study on Hydra Viridissima ………………… 203
Mussel-inspired Adhesive，Self-healing，and Injectable Poly（L-glutamic acid）/
　　Alginate-based Hydrogels ……………………………………………………… 205
高分子水凝胶整合酶的生物医学应用 ………………………………………… 208
仿生黏附可控界面 ……………………………………………………………… 209
单分子层丝素纳米带及全丝素基纳米摩擦发电机 …………………………… 209
液晶弹性体基双向形状记忆材料研究 ………………………………………… 212
纳米分子伴侣调控胰岛素的递送 ……………………………………………… 213
SAXS Studies of the Thermally-induced Fusion of Diblock Copolymer Spheres：
　　Formation of Hybrid Nanoparticles of Intermediate Size and Shape ………… 216
Molecular Pillar Approach to Construct 3D Nanomaterials for Energy Storage
　　………………………………………………………………………………………… 217
Confinement of Single Polyoxometalate Clusters in Molecular-scale Cages for
　　Improved Flexible Solid-state Supercapacitors ………………………………… 219
Lactose Targeting Photodynamic Antibacterial Materials for Pseudomonas Killed
　　………………………………………………………………………………………… 221
Self-assembled Nucleotide/Saccharide-tethering Polycation-based Nanoparticle
　　for Targeted Tumor Therapy …………………………………………………… 222
High Performance PLA Composite with Toughness and Flame Retardancy ……… 225
Light-driven Liquid Crystalline Networks and Soft Actuators with Degree-of-
　　freedom-controlled Molecular Motors ………………………………………… 227
Responsive Liquid Crystal Network Coatings and Their Applications …………… 229

Controlling the Cell Morphology in Tissue Engineering Bilayer Scaffold 232

第六部分　纤维与低维材料
Section 6　Fibers and Low-dimensional Materials

基于低维材料的人体热管理技术 239
多尺度取向半导体纤维及逻辑器件应用 240
Functional Modification and Recycling of Fiber Separator Materials 241
Functional Polyelectrolyte Materials Based on Poly(1,2,4-triazolium)s 242
Building Gradient Structure in Polymer Blend Fiber and Its Application 244
基于材料表面图案化技术揭示细胞黏附、增殖与分化的影响因素 245
可降解的金属有机纳米诊疗试剂的开发和生物应用 247
Heterogeneous Scorpionate Site in MOF: Small Molecule Binding and Activation
  248
Bio-Inspired Nanochannel Materials 249
On-surface Synthesis and Atomically Precise Fabrication of Low-dimensional
  Carbon Nanostructures 250
Preparation and Properties of Vanadium Oxides-based Electrodes for
  Supercapacitors 252
面向高比能锂离子电池的设计策略与关键材料技术 253

第七部分　新材料技术产业链
Section 7　Industrial Chain for New Materials

Plain Low Carbon Steel Resisted to Corrosion via RE Alloying 259
新型低成本耐蚀钢的应用领域及前景 261
中国金属学会团标《稀土耐候结构钢》要点解析 262
发展新型稀土耐蚀钢的产业链构建 263
德国螺纹接头正向开发对我们的启示 264
汽车紧固件产业链的思考 266
浅谈当前新形势下国内汽车紧固件供应链 268
中国金属学会团标《冷镦和冷挤压用钢》要点解析及紧固件上下游匹配 269
从原料端看汽车紧固件产业链的发展 270
汽车紧固件数字化思考 272
新冠疫情下中国汽车紧固件行业的思考 273
新形势下耐热紧固件发展的思考 274

第一部分　大会报告
Section 1　Plenary Lecture

新材料高质量发展战略研究

干 勇

国家新材料产业发展专家咨询委员会主任、钢铁研究
总院名誉院长、中国工程院原副院长

从一万年前的人类渔猎采集文明发展到今天的信息工业文明,材料的品种类型从天然的茅草木石发展到了今天的数不胜数,它就像一棵枝繁叶茂的大树一样,支撑着人类文明发展的步伐。当前我国正处于战略转型期,即从高速发展转入高质量发展阶段。面对高质量发展的需求,材料基础支撑作用不足的问题日益显现。同时,开辟新的经济增长

物化性质	作用	用途	发展阶段
金属材料	结构材料	汽车材料	先进基础材料
无机非金属材料	功能材料	电子材料	关键战略材料
高分子材料	结构功能一体化	能源材料	前沿新材料
复合材料		……	

先进基础材料	先进钢铁材料、先进有色金属材料、先进石化材料、先进建筑材料、先进轻工材料等
关键战略材料	高端装备用特种合金、高性能分离膜材料、高性能纤维基复合材料、新型能源材料、电池材料(太阳能电池材料、锂电池材料、燃料电池材料)、新一代生物医用材料、电子陶瓷和人工晶体、稀土功能材料、先进半导体材料、高纯金属、显示材料等
前沿新材料	3D打印用材料、超导材料、智能仿生与超材料、石墨烯材料、生物基合成纤维、海洋生物基纤维、蛋白质复合纤维等

点、提高环境承载能力,为我国新材料的大发展提供了难得的历史机遇。该报告分析了当前国内外新材料的相关政策,研究了2035新材料发展趋势及战略需求;梳理了先进基础材料、关键战略材料、前沿新材料等3大体系,以及新材料评价—表征—标准平台建设和军民融合重点材料,千余种重点新材料的关键技术指标,制定了2020—2035年的技术路线图,提出了新材料2035强国战略的发展路径及相关措施建议。

创新生物材料:支撑人类医疗方式变革

王靖[2],刘昌胜[1,2,*]

1 上海大学 2 华东理工大学
* 通讯作者:liucs@ecust.edu.cn

用于诊断、治疗、修复或替换人体组织、器官或增进其功能的生物医用材料是保障人类健康、提高生命质量的重要物质基础,是重要的战略新兴材料,关系到民生幸福和国家稳定。我国目前面临严峻的人口老龄化问题,对生物材料的需求剧增,且生物材料呈现出井喷式的增长。生物材料已成为决定国家未来核心竞争力的支柱产业之一。

近十多来年生物医用材料蓬勃发展,进入了一个崭新的阶段。传统生物医用材料的时代已经过去,与生物技术相结合赋予材料生物活性,使材料能与人体组织产生可控的相互作用,通过激发或调控组织再生和重建过程中的细胞黏附、迁移、生长、分化和凋亡等生理活动,增进细胞活性或新生组织的再生功能,从而帮助机体实现组织的修复和再生,已成为未来生物医用材料科学发展的方向,并正在对生物材料产业产生越来越大的影响和效应。在新型材料支撑下,各种新型的医疗器械相继出现,包括各种植入器械、心脏扩张导管和球囊、新型的药物载体以及以组织工程为代表的组织再生新策略。越来越多的材料生物学新效应正逐渐被发现,并对组织修复、诊疗等领域及临床技术产生深远的影响。例如,通过钙磷盐的体内外转化机制,可实现自固化磷酸钙人工骨材料准确塑形、生物相容、降解吸收、可注射等特性的统一,其已成功应用于骨修复及椎体成型,并推动了椎体成型微创治疗技术的发展;以骨形态发生蛋白为代表的生长因子在可控制备、高效固载和缓控释技术的突破,为大段组织损伤患者带来了福音,使原本不愈合或延迟愈合的患者能正常愈合;利用生物活性人工骨技术,诱导侧支循环血供恢复正常,可对股骨头坏死患者进行早期干预,避免关节置换,大大改善患者的生活质量。这些新材料的研发都推动了临床治疗新技术的进步,提高了医疗水平。

同时,应该看到,大多数材料临床应用远未达到令人满意的效果。目前的研究基本停留在对现象的认识,对材料学与生物学关联性的研究目前基本空白,对材料为何能产生这些生物学功能的机制不清楚,对材料学参数如何调控生物学性能的规律也知之甚少。一方面,新型医用材料不断涌现；另一方面，经典的生物相容性理论正在受到挑战，已难以解释临床应用过程中的实际问题。近年来提出"材料生物学"(Materiobiology)这一新科概念，它是采用材料学的理论和方法研究生命现象、生命过程的材料学基础(图1)。随着更多的材料生物学新效应被发现，材料生物学对新型生物材料设计与制备的指导作用显得越来越重要。因此，我们认为，开展材料生物学研究，建立材料生物学理论体系，揭示材料的理化参数对材料生物学功能的影响及其规律，并集成出材料生物学效应的数据库，对新型生物材料的构建及其临床应用具有重要的指导意义。

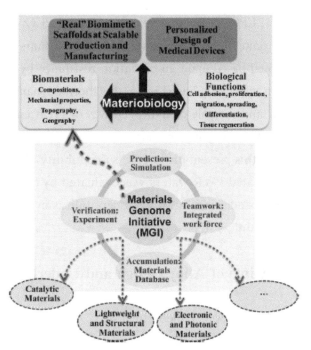

图1　材料生物学可以丰富小鼠基因表达(MGI)数据库

参考文献：

[1] Yulin Li, Yin Xiao, Changsheng Liu. The Horizon of Materiobiology: A Perspective on Material-Guided Cell Behaviors and Tissue Engineering[J]. Chemical Reviews, 2017, 117(5): 4376-4421.

Development of SCC Resistant Austenitic Stainless Steels for LWR Application with a Special Emphasis on the Role of Hydrogen

Tetsuo Shoji[1], Zhong Xiangyu[1], Xu Jian[1,2], Jiro Kuniya[1]

1 Tohoku University　2 Sun Yat-Sen University

* Corresponding author: tetsuo.shoji.c2@tohoku.ac.jp

【Introduction】

In the projects performed as OPCR at Tohoku University in collaboration with

Kansai EPCO., Tohoku EPCO., Chubu EPCO., Tokyo EPCO., Mitsubishi HI, and SN Sumitomo Metal Co., several SCC resistants 304L and 316L stainless steels were designed based upon an oxidation resistance by atomic-scale simulation by DFT. Alloy design was made based upon the thermodynamics and kinetics of metal oxidation by water and the role of hydrogen was of great concern because hydrogen can play as an oxidant as oxygen in metals.

In this presentation, SCC susceptibility of these alloys both in simulated BWR and in simulated PWR waters was evaluated by SSRT and comfirmed the effects of minor elements addition on SCC susceptibility in comparison with the reference 304L and 316L stainless steels.

【Principle of Alloy Design and Underlying Theory】

Based upon a theoretical SCC growth rate formulation based upon a crack tip strain rate model with crack growth by oxidation, EAC initiation and growth model has been proposed based upon a mechanics at crack tip and mechanisms of enhanced oxidation at crack tip as shown in Fig.1. Strain rate as well as applied stress and oxidation rate are of critical importance in SCC initiation and propagation. As can be easily seen in the modified theoretical crack growth formulation shown in Fig.1, including creep deformation, various mechanical and metallurgical factors which can influence the crack tip strain rate where hydrogen can play an important role in

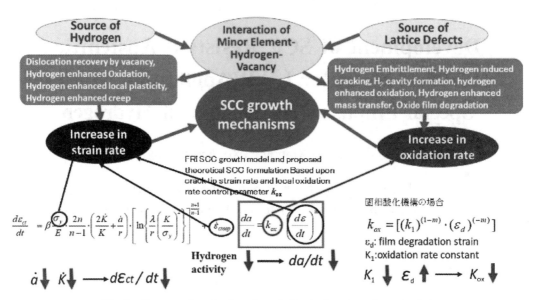

Fig. 1 Synergy of mechanics of crack tip strain rate and mechanism of crack tip oxidation — Role of Hydrogen

possible creep deformation at LWR temperature below 360℃ where interaction between hydrogen and lattice defects such as dislocation can be a cause of creep deformation at a rather lower temperature. Hydrogen/vacancy interaction is well known as a superabundant vacancy to form a hydrogen-vacancy cluster. Hydrogen also plays an important role to accelerate oxidation as well. SCC can take place as a combination of these two kinetic processes. Alloy design was aimed to reduce hydrogen activities to reduce hydrogen-induced creep and hydrogen accelerated oxidation. These interactions among lattice defects, hydrogen, and crack growth can be seen in Fig.1.

【Experimental】

Based upon the alloy design described above, several 304L and 316L based stainless steels with the addition of different minor elements such as Ce, Y, Zr, Hf, Ti, and Sc were prepared. Mechanical properties and microstructures were examined to confirm the fabricated alloys have proper mechanical properties and microstructure. Hydrogen permeability of the alloys at RT and 300℃ was measured to examine the effects of minor element addition on hydrogen diffusivity as designed. SCC resistance was evaluated by means of SSRT-SCC testing at the strain rate of 5×10^{-7}/s and 5×10^{-8}/s in simulated PWR and BWR environments. To accelerate SCC initiation and propagation, all alloys are subjected to 20% cold-work after solution annealing and machined to a smooth plate specimen with loading screw at both ends. After the tests to rupture, specimens were examined fractographically on their fracture surface and then mounted with epoxy resin and metallographically prepared to examine the number and depth of SCC crack initiation and propagation. In all tests, load-displacement curves were recorded and analyzed to evaluate the SCC susceptibility.

【Results and summary】

The SCC results obtained by SSRT were summarized and compared from a point of view of minor element addition. All alloys showed more or less SCC susceptibility in simulated PWR environments but not in the BWR environment, indicating that cold worked stainless steels are more susceptible to SCC in PWR than in BWR. SCC susceptibility evaluation by fractographic analysis suggested that Zr and Sc were found more effective to improve SCC susceptibility either in PWR and in BWR.

According to the basic SCC resistant alloy design concept as shown in Fig.1, SCC susceptibility in PWR and hydrogen permeability of the alloys are compared. Some SCC facets formed in a simulated BWR enviromnet were found at the large inclusion mainly oxides facing to specimen surface where large oxides inclusion couldn't be dispersed during the forging process from an ingot to plates due to small forging ratio. SCC facets are always associated with these inclusions. Failure of these inclusions facing surfaces can easily crack at an early stage of SSRT tests and formed crevices and local water chemistry can be developed in these crevices to promote SCC in BWR. Local acidification inside of crevices resulted in low pH and can promote SCC through hydrogen effects on crack tip mechanics in plasticity and creep, oxidation acceleration and/or hydrogen cracking. The SCC mode in the simulated PWR environment was a mixture of transgranular and intergranular but was only transgranular around cracked oxide inclusion in the simulated BWR environment.

A good correlation between SCC facets fraction on fracture surface and a total amount of Zr and Sc addition as solid solution was found as shown in Fig. 2. Increasing the concentration of Zr and Sc, less SCC facet fraction is observed. More detailed results will be presented in terms of the role of hydrogen both in crack tip mechanics and in crack tip oxidation.

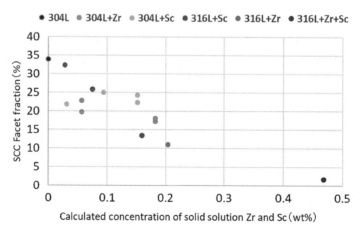

Fig. 2 Effects of concentration of Zr and Sc addition on mitigation of SCC susceptibility in 304L and 316L stainless steels

Acknowledgement Most of the work shown here was carried out in the OPCR project (2015 - 2017) and the authors would like to express their sincere thanks to the member organizations of OPCR project.

Dynamic Materials Towards Functional Adaptive Materials

Jean-Marie Lehn

ISIS, Université de Strasbourg, France

Supramolecular chemistry is intrinsically a *dynamic chemistry* because of the lability of the non-covalent interactions connecting the molecular components of a supramolecular entity and its resulting ability to exchange components. Similarly, dynamic covalent chemistry concerns molecular entities containing covalent bonds that can form and break reversibility, to allow a continuous modification in constitution by reorganization and exchange of building blocks. These features define a *Constitutional Dynamic Chemistry* (CDC) on both the molecular and supramolecular levels.

One may define *constitutional dynamic materials* as materials whose components are linked through reversible covalent or non-covalent connections and which may thus undergo constitutional variation, i.e. change in constitution by assembly/deassembly processes in a given set of conditions. Because of their intrinsic ability to exchange, incorporate and rearrange their components, they may in principle select them in response to external stimuli or environmental factors and therefore behave as *adaptive materials* of either molecular or supramolecular nature.

Applying these considerations to polymer chemistry leads to the definition of *constitutionally dynamic polymers*, Dynamers, of both molecular and supramolecular types, possessing the capacity of adaptation by association/growth/dissociation sequences. *Supramolecular materials*, in particular *supramolecular polymers* may be generated by the polyassociation of components/monomers interconnected through complementary recognition groups. *Dynamic covalent polymers* result from polycondensation via reversible chemical reactions. They may undergo modifications of their properties (mechanical, optical, etc.) via incorporation, exchange and recombination of their monomeric components. These features give access to higher levels of behavior such as healing and adaptability in response to external stimuli (heat, light, medium, chemical additives, etc.).

CDC introduces a paradigm shift into the chemistry of materials and opens new perspectives in materials science. A rich variety of novel architectures, processes and properties may be expected to result from the blending of supramolecular and molecular dynamic chemistry with materials chemistry, opening perspectives towards *adaptive materials and technologies*.

References:

[1] Lehn J M. Dynamic combinatorial chemistry and virtual combinatorial libraries[J]. Chemistry A Europear Journal, 1999, 5: 2455.

[2] Lehn J M. From supramolecular chemistry towards constitutional dynamic chemistry and adaptive chemistry[J]. Chemical Society Reviews, 2007, 36(20): 151-160.

[3] Lehn J M. Dynamers: Dynamic molecular and supramolecular polymers[J]. Australian Journal of Chemistry, 2010, 63: 611-623.

[4] Lehn J M. Chapter 1, in Constitutional Dynamic Chemistry[J]. Topics Curr. Chem, 2012, 322: 1-32.

[5] Lehn J M. Dynamers: From Supramolecular Polymers to Adaptive Dynamic Polymers[J]. Advances in Polymer Science, 2013, 261: 155-172.

[6] Lehn J M. Perspectives in Chemistry — Steps towards Complex Matter[J]. Angewandte Chemie, Internation Edition, 2013, 52(10): 2836-2850.

[7] Lehn J M. Perspectives in Chemistry — Aspects of Adaptive Chemistry and Materials[J]. Angewandte Chemie, Internation Edition, 2015, 54(1): 3276-3289.

第二部分 金属材料
Section 2　Metal Materials

体心立方金属中晶界结构的微观相场模拟研究

邱嫡[1,2],吕维洁[2],张荻[2],Metous Mrovec[4],王云志[3,*]

1 上海大学 2 上海交通大学 3 The Ohio State University
4 Interdisciplinary Centre for Advanced Materials Simulation (ICAMS)
* 通讯作者：wang.363@osu.edu

【前言】

通常晶界 5 个（宏观）自由度并不足以解释其在许多物理过程中所起的作用（如相变、塑性变形等），这是因为晶界本身还存在额外的"内部自由度"，如晶界微观结构等。我们采用结合原子模拟和微观相场（MPF）模拟的跨尺度集成模型，预测了 5 种体心立方金属中 $(1\bar{1}0)$ 扭转晶界的微观结构。该模型以原子层面的材料参数（如广义层错能面、弹性模量）为输入，确保了材料个性化特征在晶界结构演化中的影响；同时，鉴于 MPF 是一种基于能量的物理模型，能有效地消除传统几何方法在预测晶界结构方面的不确定性。研究结果表明，对于非密排的 BCC 结构金属，扭转晶界的平衡位错结构既可以呈现四边形网络结构，也可以呈现纯螺型或混合型的六边形网络结构，且该结构与材料参数紧密相关。该集成模型的建立和应用，提示我们通过采用高通量计算，可以探讨合金化方法（改变上述材料参数）对晶界结构及性质的调控作用。

【研究方法】

采用分子动力学方法计算 Nb，Mo，Ta，W 和 β-Ti 这 5 种 BCC 金属 $(1\bar{1}0)$ 晶界的广义层错能面（γ 面）和弹性模量，并将这些信息输入到相场微观模型中，模拟各金属体系中扭转晶界的动态演化过程，待体系达稳定状态后计算出晶界位错结构和类型，与传统的计算方法进行对比和分析，并探讨了材料参数对晶界微观结构的影响规律。

【结果】

1. 晶界的 MPF 方法消除了 F-B 方程在预测晶界位错结构时存在的不确定性。这是由于和 F-B 方程相比，MPF 模型中位移基矢的选择是任意的，无须是预先设定的伯氏矢量，而使用不同位移基矢所预测的平衡态晶界位错结构是唯一的。

2. 5种不同金属体系($1\bar{1}0$)扭转晶界上的位错对应的伯氏矢量是相同的,伯氏矢量共有三组,分别为 $b_1 = [001]a_\beta$, $b_2 = (1/2)[111]a_\beta$, $b_3 = (1/2)[\bar{1}\bar{1}1]a_\beta$,如图1所示。

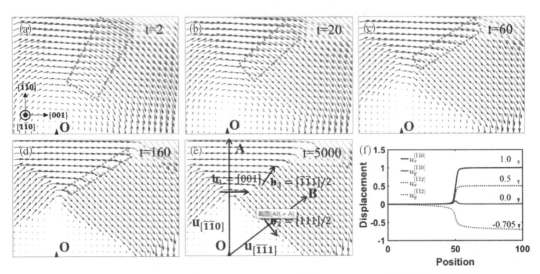

图1 位移场(a—e)和位移(f)沿 OA 和 OB 的演化($t=5\,000$,以晶格参数为单位)

3. 当体系不同时,晶界位错网络可以由混合型四边形位错网络(β-Ti)向纯螺型六边形位错网络(Mo 和 W)或混合型六边形位错网络(Nb 和 Ta)发生转变,如图2所示,且平行模拟实验结果显示目标体系的材料特性(包括弹性模量、γ 面)对晶界结构和能量的影响很大。

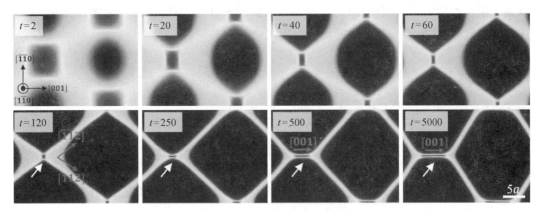

图2 钼合金在 GB 内的节点处的微观结构演变和位错反应

4. 研究结果证明了材料特性和晶界结构有着密切联系。因此,通过高通量计算,本文所提出的跨尺度的 MPF 数值模型可以进一步用来探究合金化方法(改变 γ 面和弹性模量等材料个性化参数)对晶界结构的调控作用。

Point Defects in Transparent Conducting Oxides

Qing Hou[1,*], John Buckeridge[2], You Lu[3], Thomas W Keal[3],
Alexey A Sokol[1,*], C Richard A Catlow[1,*]

1 Department of Chemistry, Kathleen Lonsdale Materials Chemistry,
University College London, London, United Kingdom 2 School of
Engineering, London South Bank University, London, United Kingdom
3 Daresbury Laboratory, Scientific Computing Department,
STFC, Daresbury, Warrington, United Kingdom
* Corresponding authors: qing.hou.16@ucl.ac.uk;
a.sokol@ucl.ac.uk; c.r.a.catlow@ucl.ac.uk

【Introduction】

For n-type transparent conducting oxides (TCO) materials such as SnO_2, In_2O_3, and ZnO, native defects play a key role in electronic conductivity. Depending on their electronic structure, energetics, and geometries, defects can act as donors, resulting in intrinsic n-type conductivity, or can compensate extrinsic donors such as Sn in In_2O_3. Predictive modeling of the properties of defects in such systems requires a detailed description of the dielectric response of the host material, which can be difficult to obtain using standard supercell techniques. Here, we employ the hybrid quantum mechanical/molecular mechanical (QM/MM) embedded cluster method, a multi-region approach that allows us to model defects at the true dilute limit, with polarisation effects described in an accurate and consistent manner. Moreover, we develop techniques to analyze the energetic balance between electrons bound to donors in diffuse and compact states, a difficult problem regardless of the model employed. We benchmark our results where possible and find good agreement with the experiment for a variety of defect-related properties.

【Computational techniques】

The hybrid QM/MM embedded cluster technique is employed to calculate bulk and defect energies in TCOs. In our QM/MM model, the inner cluster of ~100 atoms containing the central defect and its surrounding atoms is treated with a QM method, namely hybrid DFT employing the PBE0 functional. The outer region which contains ~10 000 atoms is treated with the MM method using interatomic force fields. The hybrid QM/MM embedded cluster approach is implemented in the CHEMSHELL package. The GAMESS-UK code is employed for the QM region, while the GULP package has been used to calculate the MM energy gradient.

【Results】

The calculate formation energies and charge transition levels of point defects in ZnO and In_2O_3 under Oxygen rich and poor conditions are shown in Fig. 1.

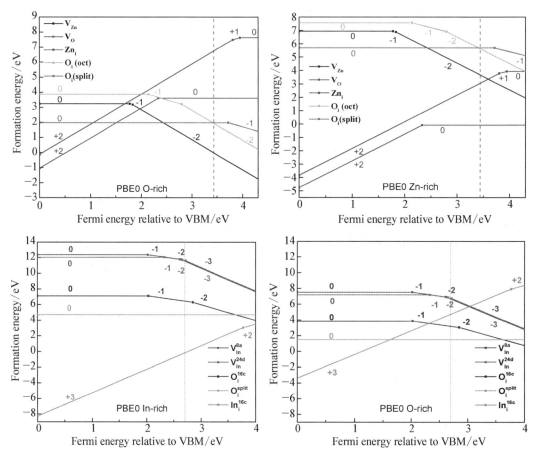

Fig. 1　Formation energy of point defects to ZnO and In_2O_3 as a function of Fermi level relative to the valence band maximum

Antibacterial Mg Alloys Through Microalloying

Gao Zhihan[1], Song Mingshi[2], Liu Rui-Liang[3], Shen Yongshuai[1], Liam Ward[2], Ivan Cole[2], Chen Xiao-Bo[2,*] Liu Xinchun[1,*]

1 Department of Orthopedics, The First Hospital of China Medical University, No. 155 Nanjing Bei Street, Heping District, Shenyang 110001, Liaoning Province, P.R. China 2 School of Engineering, RMIT University, Carlton 3053, Victoria, Australia 3 Department of Materials Science and Engineering, Monash University, Clayton 3800, Victoria, Australia
* Corresponding author: xiaobo.chen@rmti.edud.au

【Introduction】

Despite technical advancements in the design and development of new biomaterials, device-related infections continue to occur and can be life-threatening. Differing from existing research work pertains to introducing antibacterial function upon device surface, this study attempts to address such germ-infection issues through the controlled release of antibacterial species from bulk gallium (Ga) and strontium (Sr) containing magnesium (Mg) alloys.

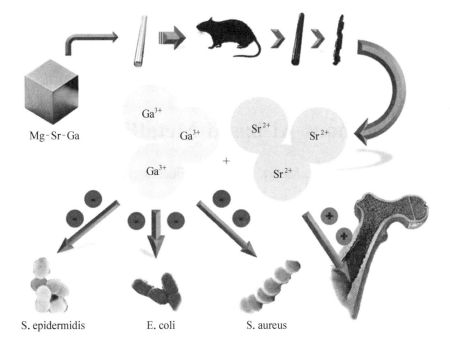

【Experimental】

To validate such a conceptual framework, Mg alloys containing micro-level concentrations of Ga and/or Sr (0.1 wt.%) were employed as model materials, along with commercially pure Mg and Ti as control groups. Biodegradation progress of such metal specimens was examined through pH and mass loss measurements, and ICP-AES as a function of immersion time in Trypticase Soy Broth (TSB) solution under physiological conditions. *In vitro* biocompatibility and antibacterial performance were examined through MTT proliferation assay with human mesenchymal stem cells (hMSCs) and the spread plate method with three representative bacterial strains, *i.e.* *S. aureus* (ATCC 43300), *E. coli* (ATCC 25922), and *S. epidermidis* (ATCC 35984). Animal tests were performed through implanting target metal rods into femurs of Sprague Dawley rats, accompanied by an injection of *S. aureus* to build a model of osteomyelitis.

【Results】

Results demonstrate that the lean addition of Ga and/or Sr reduces the degradation kinetics of the Mg matrix, and the release of Ga^{3+} ions plays a crucial role in disabling the viability of all selected bacterial strains. The histological tests confirm that the growth of fibrous tissue was accelerated in the vicinity of Mg-based implants, in comparison to that of blank and Ti controls. It is also striking that the smallest number density of *S. aureus* bacteria on the surface of the retrieved Ga-containing Mg rod implants. This proof-of-concept study provides a new and feasible strategy to address the notorious device-infection issues associated with biomedical implants for bone fracture management.

Amorphous Al-based Metallic Plastics

Gao Meng*, John H. Perepezko

University of Wisconsin-Madison

*Corresponding author: mgao63@wisc.edu

【Introduction】

Thermoplastic forming ability within the supercooled liquid region is one of the most intriguing features for amorphous materials. Novel metallic glasses (MGs) exhibit the large supercooled liquid region and the excellent mechanical properties, which enables them to behave good thermoplastic capability. Recently, a series of new MGs with a low glass transition temperature in the range of 35℃ to 150℃, called amorphous metallic plastics (AMPs), were developed, such as the Ce-, Cali-, Sr-based amorphous alloys. However, the oxidation resistance and the corrosion resistance for these systems are very poor and the cost is relatively high. Therefore, new amorphous metallic plastics with low cost and good corrosion resistance are desirable.

【Experimental】

A series of the ingots following the compositions of $Al_{92}Sm_8$, $Al_{91}Sm_8Cu_1$, $Al_{91}Sm_8Cu_2$, $Al_{91}Sm_8Ag_1$, and $Al_{91}Sm_8Ag_2$ (at.%) were produced by arc melting and then were prepared into ribbon-like samples by melt spinning method. The glassy nature of all as-spun ribbons was ascertained by X-ray diffraction. A PerkinElmer Diamond DSC and an ultrafast Flash DSC 1 were used to characterize the thermodynamic properties. A scanning electron microscope (SEM) was applied to observe the sample surface morphology.

【Results】

A new class of amorphous AlSm-based MGs were developed and were confirmed by XRD, DSC in Fig.1(a) and 1(b). All ribbons are amorphous. From Fig.1(b), it is evident that there is no glass transition signal in all heat flow curves and the primary crystallization corresponding to the precipitation of Al nanocrystals firstly appears in all AlSm-based samples marked by black arrows. Based on the advanced Flash DSC with an ultrafast heating rate of 10^4 K/s, the obvious thermal signal for glass transition can be detected and a series of glass transition temperatures for different heating rates are fitted by the Kissinger equation. Then, the glass transition temperature corresponding to the heating rate of 20 K/min can be obtained in Fig.1c. The glass transition temperatures for all AlSm-based MG systems were very close to that of some common amorphous polymers in Fig.1d, such as polyvinyl chloride (75 - 105 ℃) and were lower than the boiling point of water (100 ℃). From this point of view, the AlSm-based MG systems in this work with low glass transition temperature near the boiling point of water have the potential to become a new candidate of amorphous

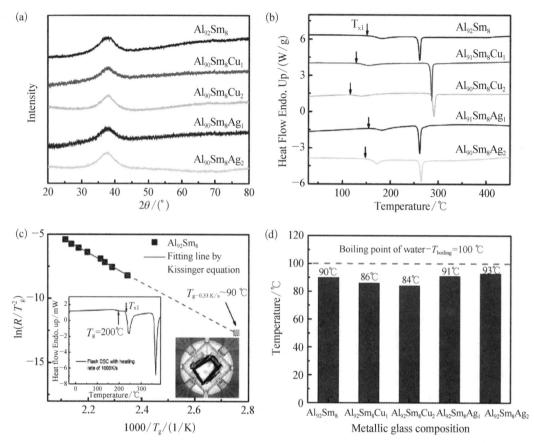

Fig. 1 (a) XRD patterns for five AlSm-based MGs. (b) DSC heat flow curves with a heating rate of 20 K/min for five AlSm-based MGs. (c) Kissinger plot for $Al_{92}Sm_8$ MG by Flash DSC. The inserted plot gives one typical heat flow curve by Flash DSC with a heating rate of 1 000 K/s and the inserted optical image shows the tiny sample is loading on the chip. (d) Histogram of the glass transition temperature for AlSm-based MGs. The red dashed horizontal line gives the boiling point of water (100 ℃)

metallic plastics.

To verify the thermoplastic processing ability for an $Al_{90}Sm_8Ag_2$ MG, an imprinting test was conducted in boiling water and the scheme of the test was shown in the left part. The bottom picture of the right part displays the distinct "UW" pattern after imprinting on the surface of the $Al_{90}Sm_8Ag_2$ ribbon in boiling water at 100 ℃ demonstrating good thermoplastic processability similar to that for conventional polymeric materials. There are no shear bands on both deformed regions, which indicates that the deformation in the boiling water is homogeneous thermoplastic deformation. These results indicate the manufacturing process including the molding, shaping, and imprinting based on the thermoplastic deformation could make AlSm-based MGs become one convenient and large-scale thermoplastic at low temperature, especially in the areas of small-scale smart devices.

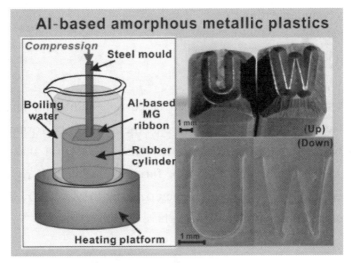

Fig. 2 One simple imprinting experiment and the impression of the "UW" pattern on the surface of the $Al_{90}Sm_8Ag_2$ ribbon in boiling water

Investigating the Compatibility of Alumina-forming Alloys with Aggressive Pb Environments for Energy Related Application

Shi Hao*, Adrian Jianu, Alfons Weisenburger, Georg Müller

Karlsruhe Institute of Technology, Institute for Pulsed Power and Microwave Technology Herman-von-Helmholtz-Platzl, 76344 Eggenstein-Leopoldshafen, Germany

*Corresponding author: hao.shi@partner.kit.edu; hao.shi@kit.edu

【Introduction】

It is well known that molten lead (Pb) and lead-based alloys (e.g. Pb-Bi eutectic) are under consideration as working fluids for energy-related applications, such as advanced nuclear reactors, concentrated solar power. However, the issue of compatibility with structural steels, in terms of corrosion and mechanical degradation, causes considerable concerns. The result of a systematic study concerning the corrosion behavior and microstructure stability of alumina-forming austenitic (AFA) alloys, during exposure to oxygen-containing molten lead and high temperature steam, is presented in this communication.

【Experimental】

Model AFA alloys have been designed based on equilibrium phase calculation, shown in Fig. 1(left). Then, the Pb corrosion experiment is performed in 10^{-6} wt.% oxygen-containing molten Pb (COSTA facility) at 600 ℃ for 1 000 – 2 000 h, and steam oxidation is tested in a horizontal tube furnace, called BOX rig, at 1 200 ℃ for 1 h. Then exposed samples have been characterized by SEM, XRD, S/TEM, XPS in terms of morphology, microstructure, and phase compositions.

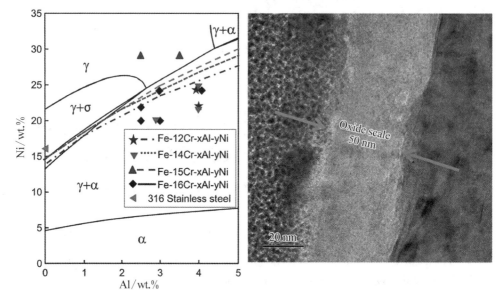

Fig. 1 Calculated phase diagrams of Fe-(12, 14, 15, 16)Cr-xAl-yNi model systems at 600 ℃ (left); TEM cross-section analysis of sample AFA54 after 2 000 h exposure to 10^{-6} wt.% oxygen-containing molten Pb at 600 ℃ (right)

【Results】

AFA alloys have shown their promising corrosion resistance to oxygen-containing molten Pb at 600 ℃. The duplex oxide scale formed on the alloy surfaces is based on an outer layer of Cr_2O_3 and an inner layer of Cr_2O_3 - Al_2O_3. In addition, a transitional layer with B2 - NiAl and Cr-rich precipitates is observed underneath the oxide scale. Regarding the microstructure, AFA alloys are able to maintain the FCC dominant alloy matrix after 1 000 – 2 000 h thermal aging at 600 ℃. Besides, some γ' -

precipitates Ni_3(Al, Fe) have been detected after the exposure test. The steam oxidation test has indicated that AFA alloys are able to forma continuous $\alpha - Al_2O_3$ scale at 1 200 ℃. The thickness of the oxide scale is around 1.5 μm. Besides, the microstructure of test samples are stable duringe.

Fatigue Behavior and Its Influencing Factors of Ti-6Al-4V Fabricated by Additive Manufacturing: An Overview

Sun Wei[1,3,*], Xi Jiangjing[2,3]

1 Shenyang University, Shenyang 110044, China   2 China Aircraft Strengten Research Institute, Xian 710065, China   3 Mechanical Engineering Department, Imperial College, London SW7 2AZ, UK

* Corresponding author: wsun@imperial.ac.uk

【Introduction】

Among the titanium alloys, Ti-6Al-4V is the most important, and it is also the most used alloy for metal additive manufacturing, accounting for about 60% of the global market. As a well-known light alloy, Ti-6Al-4V is characterized by having excellent mechanical properties and corrosion resistance combined with low weight and good biocompatibility. Therefore, this kind of material is ideal for many high-performance engineering applications, such as aerospace, high-end automobiles, and biomedical implants. Additive manufacturing (AM) is a process of generating objects from three-dimensional (3D) model data in a layer-wise manner. It can realize the rapid manufacturing of novel components and structures with complex geometries. Compared with the traditional "subtraction" manufacturing, AM has greater application value in the manufacturing of Ti-6Al-4V components because of its compressing the supply chain and reducing the waste of materials. It has been found from a comparative investigation that the mechanical properties of AMed Ti-6Al-4V are either as good as or in some cases even better than conventionally manufactured titanium alloy.

AM techniques are commonly employed to produce critical components, which usually bear cyclic loading during the whole service life. Therefore, it is a key step to understand the fatigue behavior of AMed Ti-6Al-4V, not only for the design of components but also for the evaluation of service life of them. This article provides an overview of the mechanical properties and microstructure of Ti-6Al-4V fabricated by additive manufacturing and focuses in particular on the fatigue behavior and its influencing factors.

【Experimental】

In this paper, the related research of AM Ti-6Al-4V in recent years is reviewed. By comparing the data and important achievements obtained by researchers in this field, several important problems related to AM Ti-6Al-4V are discussed, including the influence of defects, microstructure and heat treatment, residual stress, surface finish, geometry, size, and specimen orientation.

【Results】

Additive manufacturing has attracted great interest of researchers due to the significant advantages it provides over the traditional fabricationmethods. However, the design of critical load components using this technique is still at an early stage. This is mainly due to the fact that such components mainly bear cyclic load during their service life, and people have little knowledge of the fatigue performance of AMed Ti-6Al-4V compared with the components manufactured by the traditional process. Different additive manufacturing methods, the generation of defects and residual stresses, as well as the change of microstructure with process parameters, surface roughness, and anisotropic behavior, will have an important impact on the final output of AM.

(1) The defects, which are mainly composed of porosity and LOF voids, have a great influence on the fatigue properties of AMed Ti-6Al-4V. The number, size, shape, direction, and location of these defects depend on the AM process parameters, scanning strategy, construction direction, and components geometry.

(2) The extent to which fatigue properties are affected by defects depends on the ductility present in the material, so it depends on the microstructure jointly determined by the AM process and the selected heat treatment process after the AM process.

(3) In the AM process, high thermal gradient, high energy density, and rapid solidification can lead to significant residual stress and distortion of parts. The tensile residual stress will have an extremely adverse effect on the fatigue performance of

AMed components.

(4) Fatigue cracks in AMed surface conditions almost always initiate at the surface. Therefore, it will play an important role in improving the fatigue performance of AMed parts to significantly improve the surface roughness or remove it.

(5) The difference of melting speed, the time interval between layers, size, and geometry of parts will change the thermal history of the parts, which will affect the microstructure, defect composition, and subsequent mechanical properties.

(6) The orientation of the sample will affect the fatigue performance of AM parts, which will lead to anisotropic behavior.

Temperature Capability Prediction of Nickel-Based Superalloys By Cluster-Plus-Glue-Atom Model

Zhang Yu[1], Liu Bing[1], Su Zhongqian[1], You Jing[1], Sun Ye[1], Sun Xiaofeng[2], Zhang Hongyu[2], Wang Qing[3], Dong Chuang[3]*

1 Liaoning Institute of Science and Technology　2 Institute of Metal Research, Chinese Academy of Sciences　3 Dalian University of Technology

* Corresponding author: zhangjxlyu@163.com

【Introduction】

Because of higher temperature capability, nickel-based superalloys (four main kinds: (1) wrought superalloys; (2) conventional cast superalloys; (3) directional solidified superalloys; (4) single crystal superalloys from 1st generation to 6th generation) are widely used in aeronautical and stationary. However, how to prediction the temperature capability of nickel-based superalloys? Most researchers focus on the "Composition-Temperature Capability" relationship, ignoring the micro-structure of nickel-based superalloys. In order to describe the micro-structure of nickel-based superalloys, this paper built the composition formula of nickel-based superalloys with the cluster-plus-glue-atom model. A composition formula is a structural unit with three kinds of bonds: (1) Center-Shell bonds; (2) Glue-Shell bonds; (3) Shell-Shell bonds. The strengths of the bond are quantified by bond enthalpy. The composition formula is able to predict temperature capability and design compositions (7.86Co,

12.57Co, 15.71Co, 18.85Co, and 0Ta-2.65Ti) of nickel-based superalloys.

【Experimental】

The [001]-oriented single-crystal superalloys (7.86Co, 12.57Co, 15.71Co, 18.85Co, and 0Ta-2.65Ti) were produced by the selector technique. The composition was tested by XRF (X-ray fluorescence spectrometry). The standard heat treatment scheme consists of three steps: (1) 1 310 ℃/2 h + 1 315 ℃/4 h (Air Cooling, AC) for Solution treatment; (2) 1 150 ℃/4 h (AC) for first aging; (3) 1 870 ℃/24 h (AC) for second aging. The creep test was on a GWT304 - type creep machine at 1 100 ℃/137 MPa.

【Results】

Fig.1 reveals "Temperature Capability-Average Bond Strength (Bond Enthalpy)" about classic nickel-based superalloys. The relationship is as follow:

$$T_{TC}(K) = -4\,071.582 \times I_{Ave} - 1\,867.180 \quad (1)$$

$$I_{Ave} = \frac{J_{Center\text{-}Shell} + J_{Shell\text{-}Shell} + J_{Glue\text{-}Shell}}{6 \times (13 + x)} (eV/bond) \quad (2)$$

T_{TC} is Temperature Capability (K) for 1 000 hours creep rupture life at 137 MPa. I_{Ave} (eV/bond) is the average bond strength (bond enthalpy). $J_{Center\text{-}Shell}$, $J_{Shell\text{-}Shell}$, and $J_{Glue\text{-}Shell}$ are the strength of Center-Shell bonds, Shell-Shell bonds, and Glue-Shell bonds respectively. x is the number of glue atoms.

When nickel-based superalloys are rupture while the bond is breaking. As a result, the linear relationship suggests that higher-level temperature capability means stronger bond enthalpy. Every kind has its own location in Fig. 1. For instance, both temperature capability (lower than 1 180 K) and bond enthalpy (weaker than −0.75 eV/bond) are weak for wrought superalloys, such as Nimonic 80 (1 010 K), Nimonic 90 (1 050 K), and Nimonic 80A (1 100 K). Compare to wrought superalloys, conventional cast superalloys have higher temperature capability (higher than 1 180 K) and stronger bond enthalpy (larger than −0.75 eV/bond), such as TM-321 (1 260 K, −0.77 eV/bond). Removing cross bond boundary, directionally solidified superalloys have higher temperature capability and stronger bond enthalpy. For example, the temperature capability of directionally solidified Mar-M247 is about 15 K higher than that of conventional cast Mar-M247. Removing boundary strengthening

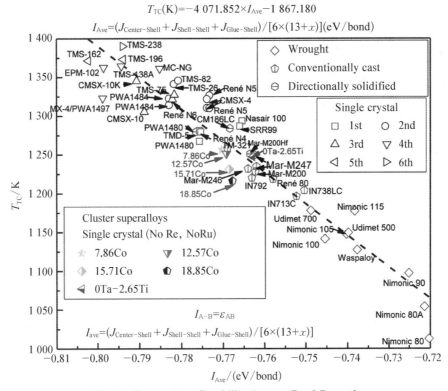

Fig. 1 Temperature Capability-Average Bond Strength

elements of directionally solidified Mar-M247, the first generation single crystal superalloys, Nasair 100 was cast. Temperature capability (Nasair 100) is increasing about 60 K. With the addition of Re and Ru elements, temperature capability is further increasing.

However, most nickel-based superalloys in Fig. 1 are mature superalloys except for cluster superalloys which are designed by the cluster-plus-glue-atom model, such as 7.86Co, 12.57Co, 15.71Co, 18.85Co, and 0Ta-2.65Ti. Cluster superalloys, the first generation single crystal superalloys, take creep test at 1 100 ℃ / 137 MPa. Transform creep rupture life into temperature capability with Larson-Miller Parameter. Fig. 1 also suggests that the cluster superalloys are also in agreement with equal (1).

Generally speaking, Fig. 1 is able to predict the temperature capability of nickel-based superalloys with the cluster-plus-glue-atom model.

The Effect of Fiber Laser Welding on Solute Segregation and Proprieties of CoCrCuFeNi High Entropy Alloy

Y. Fan[1,2,3,*], P. Li[1,*], K. Chen[1,*], L. Fu[2], A. Shan[2], Z. Chen[1]

1 China University of Mining and Technology 2 Shanghai Jiao Tong University 3 Nantong Tank Container Co., Ltd.

* Corresponding author: fanyu@cumt.edu.cn

【Introduction】

Many researchers investigated the microstructure, mechanical properties, and processing of CoCrCuFeNi, and other researchers used electron beam welding, friction stir welding, and Nd: YAG laser welding to weld different types of HEAs simultaneously. Laser welding is well established processes for joining a range of alloys. It offers a number of attractive features, such as extremely high power density, high speed, narrow weld, and small deformation In this paper, the HEA CoCrCuFeNi was prepared by using a vacuum induction furnace of intermediate frequency and then welded using a fiber laser system. The element distribution and grain refinements of high entropy alloy in welding conditions were studied. The relationship between the microstructure and mechanical properties of the welded alloy was further discussed.

【Experimental】

The equiatomic high entropy alloy CoCrCuFeNi was conducted in a vacuum induction furnace of intermediate frequency, with a water-cooled copper crucible, using CoCrCuFeNi ingots prepared from high-purity Cobalt (99.98 wt%), Chromium (99.99 wt%), Copper (99.99 wt%), Ferrum (99.95 wt%) and Nickel (99.95 wt%). Then the ingots were cut into 2 mm thick thin slices and the CoCrCuFeNi thin plates were welded by optical fiber laser. The mechanical properties of CoCrCuFeNi high entropy alloy were studied by the nano-indentation experiment. The microstructure and chemical composition were studied by XRD, SEM, EDS, and TEM.

【Results】

Combined with XRD results, the basement phase was Cu-lean FCC1 phase, hence the interdendritic phase was the Cu-rich FCC2 phase. Oval-shaped nanoparticle phases (A) with a homologous size of 20 nm were visibly discovered in Fig. 2(a), which was basically made up of Cu element.

Come to a conclusion: (1) The microstructure of the parent metal was dendrite and interdendritic, whose average grain size was about 20 μm. The microstructure of the

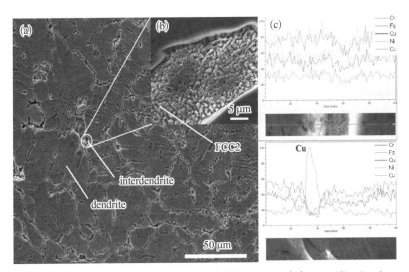

Fig. 1 Microstructure of CoCrCuFeNi: (a) SEM image; (b) Magnification image of interdendrite; (c) EDS linescan conducted in dendrite and interdendrite

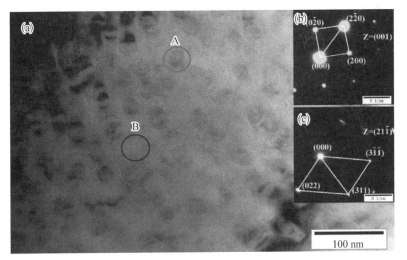

Fig. 2 TEM images of CoCrCuFeNi: (a) High-magnification bright-field image; (b) SAED image of nano-particles A; (c) SAED image of marix B

Table 1 Chemical position, crystal structure, and a fraction (in at. %) of as-cast CoCrCuFeNi HEA

Phase	Chemical composition/at. %					Crystal structure	Fraction/%
	Co	Cr	Fe	Ni	Cu		
A	4.82	5.13	2.74	13.75	73.56	FCC2	44.52
B	23.44	26.98	23.06	20.86	5.65	FCC1	55.48

fusion zone consisted of a number of equiaxed grains and directional growth of columnar crystals. Due to the fast cooling rate in the fusion zone in laser welding, the equiaxed grains of sizes were 3 to 5 μm in the fusion zone and also the columnar grains of lengths were 80 to 100 μm.; (2) Cu element was segregated in the interdendritic of the parent metal due to their small bonding energies with Fe, Co, Ni, and Cr atoms. In the fusion zone, it was noticed that Co, Cr, Cu, Fe, Ni elements were almost uniformly distributed. Grain refinement, the flow and stirring in the fusion zone during solidification made Cu element diffuse and reduced the segregation of Cu in the fusion zone; (3) The average vickers hardness of parent metal was measured to be 145.00±2.61 HV and the average vickers hardness of fusion zone was measured to be 163.62±5.56 HV. The yield stress of the parent metal and fusion zone was 207.17±23.56 MPa and 262.83±36.24 MPa, respectively. Due to the fine-grain strengthening effect, the hardness and strength of the fusion zone were greater than that of the parent metal.

The Corrosion Behaviour of Magnetocaloric Alloys La(Fe, Mn, Si)$_{13}$H$_x$ under Magnetic Field Conditions

Guo Liya*

Imperial College London

*Corresponding author: guoliya6@163.com

【Introduction】

The magnetocaloric effect (MCE)-based cooling technique is a promising alternative method to conventional refrigeration for a low carbon future *via* greater cooling efficiency as well as a reduction in the use of harmful refrigerant gases. MCE is manifest by a temperature change of magnetic material in an applied magnetic field. However, current systems are limited by the stability of the active material: in the active magnetic regenerator (AMR) cycle, the contact of the magnetocaloric alloys and the heat transfer fluid can result in severe corrosion under operating conditions. The degradation of the material might poison the giant MCE, reducing cooling efficiency, and eventually lead to failure of the refrigeration device. The limited previous work in this area has not addressed key issues associated with the coupling of magnetic field effects to the electrochemical corrosion behavior — and so do not provide appropriate insights into the *in operando* system.

【Experimental】

The degradation mechanism of the magnetocaloric refrigerants $La(Fe, Mn, Si)_{13}H_x$ in the AMR has been examined in detail. Factors representative of the real complex operating conditions within a refrigeration cycle, including magnetic field direction, the magnetic state of the material, and magneto-volume change under an alternating changing field, were all considered. Samples were moved in and out of the 1.1 T magnetic field, and the effect of the magnetic field was directly measured during electrochemical polarisation.

【Results】

The corrosion behavior can be explained by consideration of both the Lorentz and the field gradient force acting at and near to the magnetocaloric electrode surface. Whilst the Lorentz force accelerates the mass transport of reactive solution species and removal of corrosion products, the field gradient force attracts corrosion products and shows an inhibiting effect. No dramatic effects were observed under alternating field conditions suggesting that localized

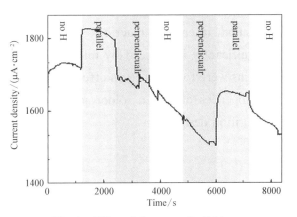

Fig. 1 **Effect of the magnetic field on the electrochemical behavior**

cracking during volume expansion does not contribute significantly to the corrosion over the short time period spanning the study. The current work suggests that for this material the magnetic field effects are largely due to magneto-transport and are most apparent when the material is in its paramagnets with a high corrosion rate (∼20 mmpy); the worst conditions are not *a priori* obvious and so conditions for accelerated tests and for applications should be carefully chosen.

Phase Formation, Transformation and Stability in Micro-alloyed Sn-based Lead-free Solder Alloys and Joints

Zeng Guang[1]*, Stuart D McDonald[2], Gu Qinfen[3], Kazuhiro Nogita[2]

1 Central South University, China 2 The University of Queensland, Australia 3 The Australian Synchrotron (ANSTO)

*Corresponding author: g.zeng@csu.edu.cn

【Abstract】

The Effect of micro-alloying additions on microstructure development of Sn-0.7 wt.% Cu during the liquid-solid transition and the stability of the dominant phases in solder joints were investigated. Solidification is a step common to nearly all soldering operations and dominant phases have the potential for crystallographic transformations and associated volume changes which may impact reliability. The nature of alloy solidification and segregation, interfacial reactions, phase formation, transformation, and stability can be significantly influenced by micro-alloying. Concurrent additions of micro-alloying elements were found to be an effective way of modifying microstructure and improving the properties of Sn-Cu solder joints, which has implications for new solder alloy composition design.

基于先进核能技术需求的特种合金材料研制

徐长征*,马天军,欧新哲,徐文亮

宝武特种冶金有限公司

* 通讯作者:xuchangzheng@baosteel.com

【引言】

　　核电是先进的清洁能源,是实现国家节能减排目标的最重要举措之一,"十三五规划"进一步明确了安全高效发展核电的思路,预计未来平均每年将建设 6 台机组左右。经过几十年采取多种形式发展核电,我国已能独立生产制造百万千瓦级核电核岛的绝大部分主设备,但某些部件的关键材料仍依赖进口,成为核电自主化建设的重要瓶颈之一。本文重点介绍了宝武特冶/宝钢特钢近年来在压水堆核电技术涉及的蒸汽发生器下封头水室隔板用镍基合金厚板、爆破阀剪切盖用镍基合金大截面棒材、屏蔽主泵屏蔽套用镍基合金冷轧薄板、镍基合金焊接材料和高温气冷堆、钍基熔盐堆涉及的镍基合金板、管、棒等关键材料研制方面取得的新进展。

【实验方法】

　　近年来,宝武特冶/宝钢特钢成功研制了蒸汽发生器下封头水室隔板用镍基合金热轧厚板,实现了 CAP1400 水室隔板全球首发;成功开发了压水堆核电站心脏——主泵屏蔽套用哈氏合金超薄板材,制造全球首卷 CAP1400 主泵屏蔽套用宽幅哈氏合金带材,打破了国外敏感物资禁运限制;成功研制了核电站非能动安全系统的关键设备——爆破阀剪切盖用超大规格镍基合金棒材,填补了国内空白;成功实现高温气冷堆示范工程蒸发器用镍基合金板、管、棒材独家整台套供货,实现了该材料的国产化,保障了工程建设。

【结果】

　　目前,宝武特冶/宝钢特钢镍基合金品种已基本覆盖目前核电在建堆型所有镍基合金牌号,产品类型从棒材拓展到钢管、热轧板、冷轧板、锻件,掌握了大型镍基合金电渣锭质量控制工艺、大规格产品均质化热加工制造工艺、大锭型坯料锻造、宽板幅超长薄钢板热轧、大规格超重热挤压钢管制造工艺、产品热处理与性能控制技术等关键核心技术,并已

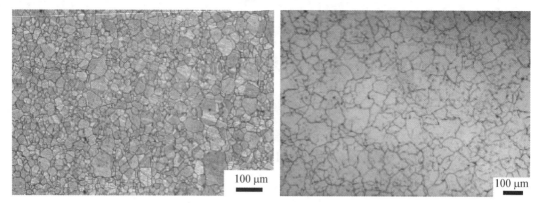

图 1 国产 690 合金厚板及大规格棒材显微组织

形成核电用镍基合金常规耐蚀、高温性能、焊接性能评价能力,所开发的产品完全符合先进核电技术规定的特殊成分要求、特殊规格/极限规格要求、性能要求、特定工况应用性能要求,产品批量进入市场并完成工程应用,为核电自主化进程和国家能源战略安全提供了有力的支撑和保障作用,具有重大的经济效益和社会意义。

Microstructure, Micro-chemistry and Stress Corrosion Cracking of Dissimilar Metal Weld 16MND5/309L/308L/Z2CND18-12N Used in Nuclear Power Plants

Li Guangfu*, Yuan Yifan, Lu Xu

Shanghai Research Institute of Materials, Shanghai Key Lab for Engineering Materials and Evaluation, Shanghai Failure Analysis & Safety Evaluation Center, Shanghai 200437, China
* Corresponding author: liguangfu@srim.com.cn

【Introduction】

In the nuclear island of a pressurized water reactor nuclear power plant, low alloy steel (LAS) such as A508III or 16MND5 is used to make reactor pressure vessel and stainless steel (SS) such as 316L or Z2CND18-12N is used for making piping. The connection between a pressure vessel and piping is a dissimilar metal weld made by using either of two kinds of filler metals: (1) nickel-based alloys; (2) SS such as 309L/308L. Since 2000, many stress corrosion cracking (SCC) incidents occurred in the nickel-based welds of Swedish, USA, and Japan power plants, which have raised the concern in the integrity of the dissimilar metal welds in service significantly. Many kinds of researches have been performed on the SCC behaviors of the nickel-based welds. However, not much work has been published on another kind of weld which is made by using SS filler metals such as 309L/308L.

[Experimental]

The microstructures and chemical composition profiles of the dissimilar metal weld 16MND5/309L/308L/Z2CND18-12N weld were examined with microscopies and energy dispersive X-ray analysis. Hardness distributions were measured via a diamond-pyramid hardness machine using a load of 500 g. The SCC property of the weld in simulated PWR primary water at 290 ℃ was characterized by using slow strain rate testing (SSRT) on parallel tensile specimens covering all parts of the weld, i.e. from the 16MND5 base metal to Z2CND18-12N base metal. An external Ag/AgCl reference electrode and a potentiostat were used to measure/control the electrode potential. In the present paper, all the potentials are quoted with respect to the standard hydrogen electrode (SHE).

[Results and discussion]

The microstructure and micro-chemical composition profiles across the transition zone of the weld are shown in Fig.1. There was a thin martensitic layer along the fusion line. The transition zone was an austenitic layer, in which the gain boundaries mainly perpendicular to the fusion line are named as Type I and those parallel to the line are Type II. The hardness peak of the weld was at the 16MND5/309L fusion line.

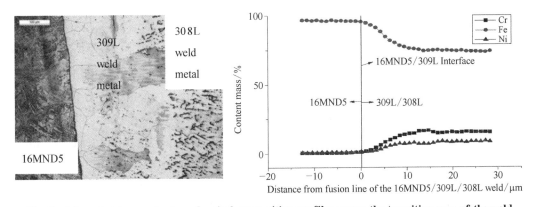

Fig. 1 **Microstructure and micro-chemical composition profiles across the transition zone of the weld**

SSRT tests for SCC in simulated PWR primary water environment at 290 ℃ were performed at different electrode potentials in the range from -720 to $+200$ mV (SHE) that simulated the electrochemical conditions of the weld in the environments from ideal water chemistry to bad water chemistry with significant contamination of oxygen. No apparent evidence of SCC was found on the specimens tested at lower potentials from -720 to $+100$ mV(SHE), which all failed by a ductile fracture in 308L bulk weld metal during SSRT, similar to that tested in inert gas N_2. The specimen tested at $+200$ mV failed by brittle SCC in the area around the 16MND5/309L interface. The SCC cracks were transgranular both in 16MND5 HAZ and at the interface, but were mainly intergranular in the austenitic layer of the transition zone of 309L weld metal, as shown in Fig. 2.

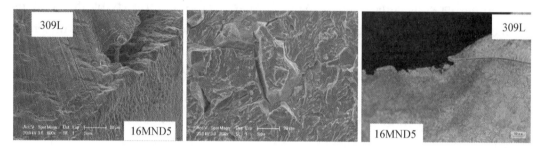

Fig. 2 Fracture surface and section of SCC around the fusion line tested at $+200$ mV(SHE) in the water at 290℃

The test results revealed that the LAS and SS base metals, the bulk weld metal and its interface area closed to the SS base metal had relatively much higher resistance to SCC, but the area around 16MND5/309L interface was the weakest part of the dissimilar metal weld. Results showed that SCC only occurred at high potential, which means bad water chemistry with significant contamination of oxygen in the real plants. This can be explained by using both the Slip-Dissolution Model for SCC in high-temperature water environments, and the research results on the relationship among water chemistries in bulk solution and crack tip solution and SCC behavior of metals in the environments. According to the model, SCC crack growth is realized through the repetition of the following steps: (1) mechanical rupture of protective oxide film due to straining at the crack tip, which produces bare surface; (2) anodic dissolution of the bare surface, leading to crack growth; (3) film formation again on the bare surface, showing repassivation. When the metal is exposed to high-temperature water containing high content of dissolved oxygen or at a high potential, the bulk surface including the crack mouth exhibits high potential but the crack tip is always at low potential due to the rapid consumption of oxygen inside. Therefore a potential gradient exists from the crack mouth to tip, which causes harmful anions like Cl⁻ to get enriched at the tip leading to decreased repassivation ability of the material at the tip, thus causing localized anodic dissolution there, i.e. SCC cracks growth.

【Conclusions】

The interface areas around 16MND5/309L and 308L/Z2CND18-12N exhibited apparent fusion lines with complicated microstructure/chemical composition changes. There was a thin martensitic layer along the fusion line, which possessed the highest micro-hardness among the weld. The transition zone was an austenitic layer within which Cr% and Ni% changed significantly. The area around 16MND5/309L was the most susceptible to SCC among the weld in the simulated PWR primary water environment. Both intergranular and transgranular SCC occurred in this area when tested in at high electrode potential which is mainly related to bad water chemistry with high oxygen contamination.

钢铁材料性能极限化探讨与初步结果：强度、塑性、韧性

曹文全*，王存宇，
俞峰，翁宇庆

钢铁研究总院特钢所，北京，100081
* 通讯作者：caowenquan@nercast.com

【前言】

航空航天、交通运输、海工装备、矿山机械和无线通信等领域高端装备的发展需要高强塑积钢、超高强韧钢以及结构性能一体化材料。高端装备对更高性能钢的需求，为钢铁材料性能极限化、钢铁材料强塑韧倒置关系以及结构性能一体化研究提供了契机。本文主要针对钢铁材料的强度、塑性、韧性、低密度和长寿命等性能极限化研究，介绍了高强度、高塑性以及高韧性钢的亚稳化、层片化和多尺度化的发展思路与部分研究成果。

【研究方法】

利用热力学软件和理论计算，通过中锰合金化设计和微观组织结构调控，进行了高强

韧塑钢研究；通过中高合金含量成分设计和组织调控，进行了超高强度钢研究；通过高氮和高铝合金化设计，进行了高耐蚀高硬度轴承钢和低密度高强塑韧钢研究；通过超细化与均匀化组织调控，进行了超长寿命轴承钢研究。

【结果】

通过对中锰钢汽车钢、低合金超高强钢、高氮合金化与高铝合金化研究与相关微观组织性能调控研究，获得了抗拉强度 750 MPa 和最高延伸率 165% 的突破性研究成果；获得了 $R_{p0.2}=1.7\sim2.1$ MPa、$R_m=2.0\sim2.5$ GPa 和 A_5 约 10% 的低合金超高强度钢，$R_{p0.2}=2.0\sim2.2$ GPa、$R_m=2.3\sim2.5$ GPa 和 A_5 约 10% 的高合金超高强钢和 $R_m=2.8$ GPa 和 A5 约 5% 等研究成果；获得了硬度不低于 58 HRC 的高耐蚀高氮钢和密度约 6.5 g/cm³ 的低密度钢和接触疲劳寿命 $L_{10} \geqslant 1\times10^8$ 次的突破性进展。上述研究结果表明，通过科学的合金化设计，可行的组织性能调控和钢铁材料性能极限化的研究，可以突破传统强度、塑性和韧性倒置关系，实现钢铁材料性能的跨越式发展，满足高端装备对更高性能钢铁材料的需求。

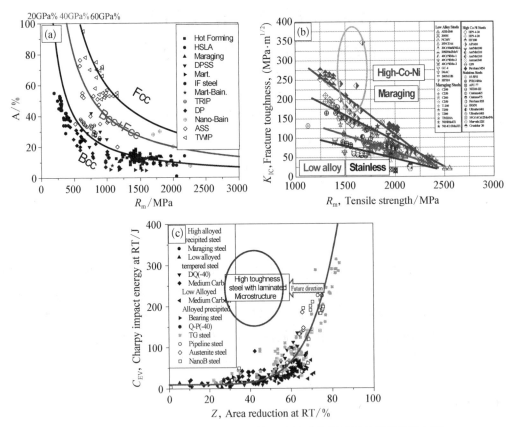

图 1　钢铁材料的塑性和韧性随着抗拉强度的升高而降低的强韧塑性倒置关系(a)和(b)和解决钢铁材料强塑韧倒置关系的可能方向(c)

Nb Microalloyed Ultrafine High C Spring Strip and Its 50 Year Evolution

Yin Jiang*

Jiangyin Xingcheng Special Steel Co., Ltd

* Corresponding author: yinjiang@asac.cn

【Introduction】

The high carbon spring steel strip 62Si2Mn developed by Shanghai fifth steel mills in the 1960s belongs to the most widely used spring steel. The manufacture of high carbon steel strip for coil spring used to go through more than 25 processes such as billet-hot rolling-repeated cold rolling and protective atmosphere annealing-quenching-tempering. Shanghai clock material factory tried to increase single pass cold rolling compressibility ε, from the original ≤0.35 increased to 0.73, to reduce the process and improve the performance of the strip. however, although the effect is obvious, the breaking rate of the strip also fierce increase. (Fragments flew like bullets into the veins of the bare arms of the regulators and workers, retain all their life) As a supplier, it is urgent to confirm the possibility of increasing the cold rolling compressibility of high-carbon spring steel, which is unprecedented at home and abroad. In the late 1970s, the author has part-time management of the steel products, in the 1980s, Imao Tamura put forward the theory of controlled rolling. Follow this train of thought, based on the actual rate range of thermal deformation, the relation of the accumulative compression rate with Zener Hollomon parameter Z was researched and the rules of $\log Z = 29.4 \sum \varepsilon - 0.78$ was mastered and a new technology design was applied successfully on the traditional production line by author, at the same time, the size tolerance of the finished products was narrowed. Since the thermal deformation austenite is in the dynamic recrystallization interval, a fine equilibrium microstructure ferrite-pearlite can be obtained in the subsequent controlled cooling transition. The lamellar spacing of the pearlite <0.3 μm, so the maximum allowable cold rolling ε is 0.8, cold rolling and annealing trace number can be reduced by more than 50%. In the 1990s, when the user equipment was reformed, the previous short folded strip had to be improved into a long coil strip. The cooling rate in the long coil strip is <1 K/s slower than $4\sim7$ K/s of shorter folded strip, continue to use the above process, austenite into coarse ferrite and lamellar spacing of pearlite greater than 1.2 μm, prone to appear bainite microstructure if the cooling is rapid, performance also deterioration. On the side, the large-scale production of Nb microalloyed high carbon spring steel is reported rarely.

【Technological innovation and technological economic benefits】

Microalloyed steel technology was published in 1975. According to this, the author has researched and produced 62Si2Mn steel with >0.005 wt% Nb. The CCT curve of the hot deformation austenite with ASTM G No 10-grade fine grain was moved up and left after controlled rolling. The equilibrium transition temperature $\geqslant 973$ K, that is, the phase transition temperature of the A-F-P was increased by more than 50 ℃, the fine lamellar spacing of pearlite is $\leqslant 0.2$ μm (Fig. 1) and subsequent coiling is completed at $\geqslant 923$ K, namely, the process window is improved. To avoid the formation of coarse pearlite during the phase transition of the coiling strip in slow cooling, and to reduce the decarburization of the surface, thus ensuring the implementation of the original cold rolling process. The performance of the product continues to be stable. For example, the failure rate drops from 20% to it goes to zero when the coil spring was repeated loosening and frapping for 3 000 times. (The product was awarded the silver medal of product quality of the People's Republic of China three times from 1985 to 1995). Nb addition leads to an increase in the temperature of dynamic recrystallization in the thermal deformation zone of the austenite of 62Si2Mn, and the fine austenite increases the phase change temperature in the

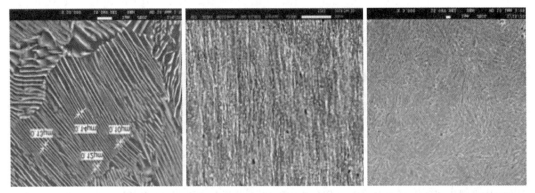

Fig. 1 Ultrafine microstructure, SEM Left, lamellar spacing $\leqslant 0.14$ μm as hot rolled and slow cooling; middle, as ε 0.8 cold rolled; right, original austenite grain diameter 7 μm, the substructure is still finer as Q-T

continuous cooling process, and then a fine F-P is formed which is independent of the cooling rate after coiling, namely, the process window is improved. The trademark Super6 strip steel was created by the author as a field engineer aims to surpass JIS SUP6 (i. e. 60Si2Mn). Adjusting internal control components and process parameters according to the similarities and differences of the production line, it has been successively produced in Shanghai fifth steel mills, Baosteel special steel, Jiangmei spring, Baosteel Co., Ltd. Nanjing steel (from 2012) and other enterprises. Spring steel and its strip still get ultrafine grain on the improved traditional production line (Fig. 1). The strip with ultra-high tensile strength 1 800 MPa has high-cost performance and has a wide application range from aerospace to home. For example, as a car seat belt spring (Fig.2), the German kern-liebers CHINA company uses 5 000 tons of Super6 strip to make coil spring every year. CHINA's automobile safety belt products were about 18 million sets in 2012 and the company owned 70% of the seat belt market share from 2016.

Fig. 2　Super6 coiling spring. left, coining spring 7.8 g; middle, 1 round coin; right, car seat belt spring

【Strategic thinking on the development approach of national iron and steel technology】

Although the examples in this paper are a kind of products developed by enterprises, they also involve the direction and approach of various research. Large deformation involves in austenite deformation texture and evaluate whether performance requires application of texture analysis method for this example, due to late phase transformation, the texture is not easy to observe and test, the answer is no as an enterprise site engineer. First of all, must pay attention to the process and the use of macro phenomena such as performance, avoid dig in texture and variation, and other advanced theory. The authors also look forward to the mature of advanced technology, that has forecast function in the future.

Microstructure Evolution and Crack Propagation of Pearlitic Wheel-rail Steel under Service Condition

Chen Hu, Zhang Chi*, Chen Hao, Yang Zhigang

Key Laboratory of Advanced Materials of Ministry of Education, School of Materials Science and Engineering, Tsinghua University, Beijing 100084, China

*Corresponding author: chizhang@tsinghua.edu.cn

【Introduction】

The initial microstructure and its evolution during service determine the fatigue properties of materials. However, the mechanism of microstructure evolution of rail-wheel steels under the wheel-rail contact loading condition and its influence on the fatigue resistance is still controversial. In this paper, the characterization of microstructure and performance evolution, the simulation of microstructure evolution, and the prediction of crack growth of wheel-rail materials under service conditions are systematically studied. The predictive model of crack growth based on the microstructure evolution of wheel-rail material is further formulated, which may provide an experimental and theoretical basis for the design of wheel-rail materials.

【Experimental】

In this paper, the structure evolution and crack propagation of the contact surface of the wheel and rail are studied by means of experimental characterization and numerical simulation. By constructing an elastic-plastic phase-field model, a multi-field model is proposed to quantitatively describe the evolution of cementite in pearlite under rolling contact. The coupling of microstructure evolution and crack growth was realized by introducing crack field and microstructure-related fracture toughness parameters, and the crack growth behavior based on microstructure evolution under rolling contact condition was simulated.

[Results]

The rolling contact fatigue testing machine is employed to simulate the wheel and rail contact. Experimental results show that the pearlite structure of the surface layer in the rolling-sliding sample transformed into nanocrystalline Fe-C alloy in which cementite underwent full or partial decomposition while the pearlite lamellae appeared unperturbed in the pure rolling sample. The effect of contact temperature rise is eliminated through the design of pure rolling and rolling-sliding contact contrast experiment. The white contrast layer in the rolling-sliding sample is formed by cyclic plastic shear deformation rather than frictional heat-induced austenitization and subsequent martensitic transformation. The hardness of pure rolling and rolling-sliding contact surface has a different distribution, consistent with a the distribution of the corresponding maximum shear stress, which further proves the deformation mechanism of the formation of the white contrast layer.

A phase-field model coupled with a finite element model and plastic strain accumulation model is originally proposed to simulate the real-time microstructure evolution subjected to rolling-sliding loading, which can well predict the evolution of cementite morphology, cementite volume fraction, and carbon distribution. Upon experimental validations, the proposed model predicts more accurate and realistic results than Sauvage's model. Three-stage dissolution kinetics of cementite lamellae is revealed from the phase-field simulations. Due to the kinetics transition, a great microstructure gradient is predicted along with the contact depth, which well explains the experimental observation. Besides, the effect of elastic strain-induced free energy contribution and ferrite/cementite interface thickness on cementite

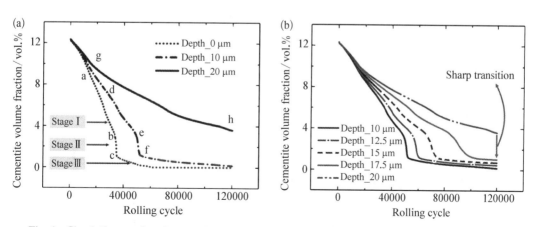

Fig. 1 Simulation results of cementite dissolution kinetics at different contact depths: (a) Curve of three-stage dissolution kinetics; (b) Dissolution kinetics of cementite near the transition zone

dissolution kinetics is studied.

A phase-field model, which is validated by comparing the simulated critical deflection angles with the results from theoretical analysis of the linear elastic fracture mechanics and the experiments, is formulated to study the crack propagation in layer materials and polycrystalline materials. Through the coupling of microstructure field and crack field, a microstructure-evolution-dependent crack propagation phase-field model under rolling contact condition is proposed, which can be used to study the influence of the initial microstructure on crack propagation resistance. It is found that the crack propagation resistance increases with the decrease of initial pearlite lamellar spacing, which is consistent with the experimental results.

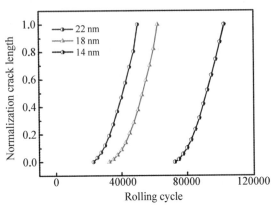

Fig. 2 The change of crack length with rolling cycle under different initial cementite layer thickness

双相钢损伤形核的介观起源

唐骜[1], 刘海亭[1], 刘桂森[1], 赖庆全[2], 钟勇[3],
王利[3], 王建峰[4], 卢琦[4], 沈耀[1]*

1 上海交通大学材料科学与工程学院金属基复合材料重点实验室
2 南京理工大学格莱特研究所　3 宝钢汽车用钢开发与应用技术国家重点实验室　4 通用汽车中国科学院
* 通讯作者：yaoshen@sjtu.edu.cn

【前言】

受到节能减排的驱动,汽车工业对先进高强钢(Advanced High Strength Steel, AHSS)的需求日益增加。在商业 AHSS 中,双相钢(Dual Phase, DP)因其高强度以及良好的成型性使其应用最为广泛。双相钢强度和塑性的良好结合源于其复合材料类型的微观结构,即包含承担应力的硬相马氏体和承担应变的软相铁素体。然而,这种两相性质的巨大差异会带来变形时的应力或者应变集中导致损伤形核和过早断裂。本文研究双相钢损伤形核的模式并通过识别各模式的介观应变和应力分布特征鉴别了不同损伤模式的介观起源,以期为双相钢微观结构设计实现损伤抑制提供参考。

【研究方法】

本研究以具有相近的铁素体强度和马氏体强度,但马氏体含量和分布不同的三种双相钢为原材料,采用原位拉伸结合 SEM 的方法原位观察损伤的形核部位并统计不同模式损伤的数量。进一步,采用 SEM 下的数字图像相关方法(μ-DIC)表征应变分布特征。该方法能够获得样品变形后表面的应变分布图像,并已广泛应用于双相钢局部应变非均匀性的研究。采用基于微观结构的有限元计算表征应力分布的特征。最后,综合分析不同模式损伤形核部位的应力、应变、应变梯度特征,揭示损伤形核的介观机制。

【研究结果】

结果表明,随着马氏体体积分数(V_m)的增加,主导的损伤形核模式由铁素体晶界(F/F)脱粘转变为铁素体/马氏体界面(F/M)脱粘,最终转变为马氏体开裂(图1)。虽然所有的形核模式都可能是由原子尺度上的应力集中引起的,但本研究显示它们在介观尺度上有着不同的起源:(1) F/F 的脱粘主要是由高介观尺度应变引起的晶界附近位错堆积以及晶界位错残留导致的。(2) F/M 脱粘是由高介观尺度应变梯度引起的位错堆积导致的。(3) 马氏体开裂是由马氏体中的高介观尺度应力引起的。这些研究结果表明,损伤的形核模式直接受介观尺度应变和应力分布特征的控制,而介观应变和应力分布特征又主要取决于微观组织,如本文中的马氏体体积分数和分布。因此,控制介观尺度应变和应力分布是调控显微组织以抑制损伤形核进而提高双相钢等复合材料类型韧性合金抗损伤能力的核心。

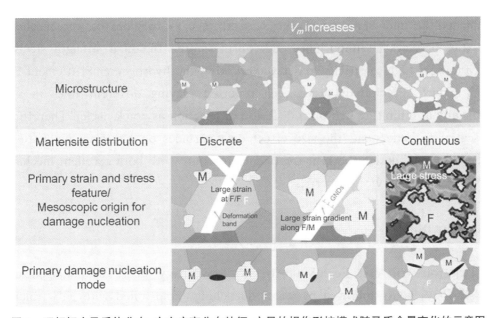

图 1 双相钢中马氏体分布、应力应变分布特征、主导的损伤形核模式随马氏含量变化的示意图

The Size Effect of κ-carbides Precipitation on Mechanical and Hydrogen Embrittlement Properties of a Fe-Mn-Al-C Low Density Steel

Guo Xiaofei[1*], Marta Lipińska-Chwałek[2], Wolfgang Bleck[1]

1 Steel Institute, RWTH Aachen University, Germany 2 Central Facility for Electron Microscopy, RWTH Aachen University; Ernst Ruska-Centre for Microscopy and Spectroscopy, Forschungszentrum Jülich GmbH Ernst Ruska

* Corresponding author: xiaofei.guo@iehk.rwth-aachen.de

【Introduction】

The low-density steels of the Fe-Mn-Al-C system have received much attention on the combination of high strength (∼1 000 MPa) and superior ductility (∼50%), exhibiting great application potentials for the lightweight infrastructures. A unique characteristic of this steel grade is the formation of κ-carbides, which is the ordered face-centered cubic phase of $(Fe, Mn)_3AlC$, upon low-temperature annealing. The κ-carbides strongly raise the yield strength of the materials with maintaining good ductility. Since there is a large space for tailoring the yield strength by controlled heat treatment, the understanding of the size effect of κ-carbides on the mechanical properties becomes an important issue. Furthermore, the high strength steels are known to have high sensitivity to delayed cracking and hydrogen embrittlement (HE) during service. Precipitates have been reported playing different roles in either beneficially as inhibiting hydrogen diffusion or function as crack nuclei. Therefore, it is essential to understand the size effect of κ-carbides on hydrogen embrittlement susceptibility and design the heat treatment procedure with both excellent mechanical properties and in-service properties.

【Experimental】

Twin-roll strip cast Fe-30Mn-8Al-1.2C low-density steel was homogenized at 1 150 ℃ for 5 hours followed by water quenching, cold-rolled with 50% thickness reduction from 2.2 mm to 1.1 mm, recrystallization annealed at 900 ℃ for 1 hour with water quenching, and further heat-treated at 600 ℃ for 1, 5, 15, and 60 min respectively with water quenching to induce κ-carbides with different sizes. The microstructures of the investigated heat-treated specimens were investigated by LOM, TEM. The mechanical properties were characterized by quasi-static tensile tests at $10^{-3}s^{-1}$, micro-hardness test, and nano-indentation test. The HE susceptibility was characterized by slow strain rate tests at $10^{-6}s^{-1}$ with hydrogen cathodic pre-charging. The hydrogen trapping behavior was identified by thermal desorption analysis (TDA). The fracture surfaces were further examined under SEM to reveal the degradation mechanisms.

【Results】

The mechanical properties of the heat-treated specimens reveal gradually increased yield strength from 551 MPa to 953 MPa as the heat treatment period has been extended from 0 min to 60 min, in Fig.1(a). The yield strength immediately raises by 223 MPa after 1 min annealing. The ductility remains stable at the range of 52% to 57% till 5 min annealing period, then declined to 30% after 15 min annealing and further to 20% after 60 min annealing. At all the heat-treated conditions, the specimens reveal austenite-based phase with the average grain size of ~20 μm, Fig.1(b). The HR-TEM revealed the ultrafine granular κ-carbides in the size of 2 - 4 nm at the 600 ℃ 1 min condition and the cuboidal κ-carbides in the size of ~10 nm at the 600 ℃ 60 min condition. The steep raising of yield strength was attributed to blocking dislocation gliding by Orowan bypassing of the nanoscale κ-carbides. Also, the shearing of κ-carbides is expected in the late deformation phase, especially in annealed specimens with a large size of κ-carbides. In parallel with the growth of κ-carbides, the extended annealing time leads to the formation of filmy carbides at the grain boundaries, as revealed in the LOM images in Fig. 1 (c). The fracture surface observation reveals that grain boundary decohesion occurs at the early deformation stage due to the grain boundary carbides. The introduction of hydrogen further facilitates the decohesion process and leads to abrupt mechanical degradation shortly after the yield point, as shown in Fig.1(d). Therefore, the grain boundary carbides restrain the material plasticity and strongly deteriorate the HE resistance. In comparison, the rapid precipitated intragranular k-carbides (2 - 4 nm) at a very short annealing periods greatly increase the yield strength and function as effective hydrogen traps. The thermal desorption analysis gives further information about the capability and revisability of hydrogen in the κ-carbides with different sizes.

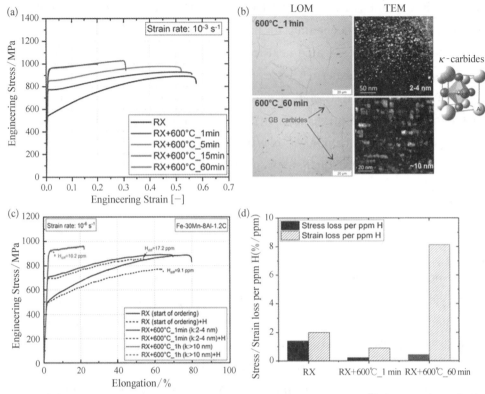

Fig. 1 (a) Tensile properties of Fe-30Mn-8Al-1.2C after heat treatment at 600℃ from 1 min to 60 min; (b) Microstructure after LOM and HR-TEM revealing austenite-based microstructure with different sizes of κ-carbides at 600℃ 1 min and 60 min conditions; (c) Hydrogen embrittlement susceptibility characterized by slow strain rate tests and (d) the evaluation indexes of stress loss and strain loss

含2%Mn多相组织(M3)低合金钢中纳米Cu析出极限强化研究

尚成嘉[1,3,4],韩刚[1,2]

1 北京科技大学钢铁共性技术协同创新中心 2 北京科技大学新材料技术研究院 3 海洋装备用金属材料及其应用国家重点实验室 4 上海大学材料科学与工程学院

【前言】

具有多相组织(M³)特征的第三代低合金钢通过多步亚临界热处理[1-17],调控板条马氏体/贝氏体基体中的残余奥氏体和Cu等纳米析出[13, 15-17],使其具有高强度、高韧性和高塑性,是一类具有高强度、更高服役安全的低合金钢。残余奥氏体对韧塑性均有重要贡献,亚临界处理引起的马氏体/贝氏体基体的软化可以通过NbC、Cu的析出相得以强化。本文介绍了Cu在多步亚临界热处理过程中的演变及其对多相组织(M^3)低合金钢的强化效果,利用具有B2结构的纳米Cu团簇可实现极限强化,多相组织低合金钢的强度达到了1 000 MPa级,韧塑性优异、屈强比低。

【实验过程】

实验钢成分为 Fe-0.087C-0.79Si-1.041Al-2.05Mn-2.12Cu-1.50Ni-0.25Mo-0.16Nb (wt.%),热轧直接淬火,钢板厚度为8 mm。实验钢 Ac_1 和 Ac_3 温度分别为690 ℃和943 ℃。经过780 ℃保温30 min一步临界热处理的钢,直接淬火至室温。得到的实验钢记作 IA 30。IA 30 钢的奥氏体相变点起始温度 Ac_1' 为675 ℃,完全奥氏体化温度 Ac_3' 为948 ℃。第二步临界回火温度选择略高于 Ac_1' 温度的680 ℃。而第二步临界回火的时间则选取了4个不同的时间,分别为5 min、15 min、60 min、180 min。经过不同临界回火时间热处理的实验钢分别记作 IA 30+IT 5、IA 30+IT 15、IA 30+IT 60、IA 30+IT 180。

【实验结果】

利用三维原子探针和高分辨透射电镜,揭示了Cu两步临界热处理过程中析出相的形态、尺寸、成分、析出密度、析出相的体积分数、晶体结构、析出相与基体的取向关系及错配程度随保温时间的变化规律。第二步回火后,随回火时间由短变长,Cu析出晶体结构的演化序列为:B2 FeCu→9R Cu→退孪晶9R Cu→复杂过渡相→FCC Cu。析出相与基体的取向关系、错配度:B2 FeCu相与基体的取向关系为$(1 1 0)_{B2}//(1 1 0)_α$,$[0 0 1]_{B2}//[0 0 1]_α$,为共格析出相;9R Cu 与 α-Fe 基体的取向关系为$(1 1 -4)_{9R}//(0 1 1)_α$,$[-1 1 0]_{9R}//[1 -1 1]_α$,为共格析出相。IA 30+IT 5是综合力学性能最佳的实验钢,其对应的Cu析出相为2~3 nm的B2有序相(图1),其屈服强度接近1 000 MPa,总延伸

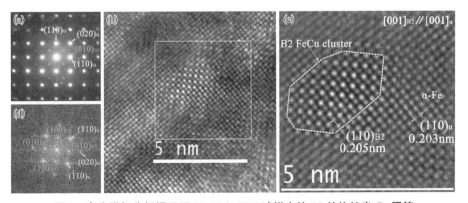

图1 高分辨相分析揭示了 IA 30+IT 5 试样中的 B2 结构纳米 Cu 团簇

率达到22.5%(图2)。场发射透射电子显微镜证实了 IA 30＋IT 5 钢中有纳米 γ,它出现在回火马氏体和临界铁素体的板条界面上。瞬时加工硬化曲线表明纳米 γ 增强了钢材的塑性。尽管 IA 30 ＋ IT 5 钢中逆转变奥氏体的体积分数和尺寸都非常小,但还是为钢材提供了一定程度的加工硬化效果[图 2(a)]。

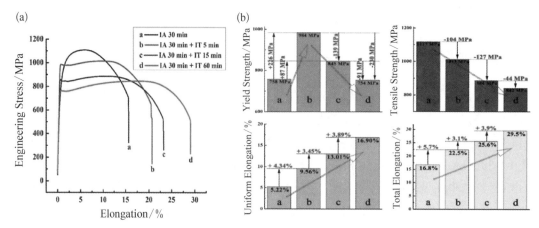

图 2　不同工艺得到的 Cu 析出相的析出强化效果对比

参考文献:

[1] W.H. Zhou, H. Guo, Z.J. Xie, C.J. Shang, R.D.K. Misra. Copper precipitation and its impact on mechanical properties in a low carbon microalloyed steel processed by a three-step heat treatment, Mater. Des.[J], 2014, 63: 42 - 49.

[2] W.H. Zhou, X.L. Wang, P.K.C. Venkatsury, H. Guo, C.J. Shang, R.D.K. Misra. Structure-mechanical property relationship in a high strength low carbon alloy steel processed by two-step intercritical annealing and intercritical tempering[J], Mater. Sci. Eng. A, 2014, 607: 569 - 577.

[3] Wenhao Zhou, Hui Guo, Zhenjia Xie, Xuemin Wang, Chengjia Shang. High strength low-carbon alloyed steel with good ductility by combining the retained austenite and nano-sized precipitates [J], Mater. Sci. Eng. A, 2013, 587: 365 - 371.

[4] W.H. Zhou, V.S.A. Challa, H. Guo, C.J. Shang, R.D.K. Misra. Structure-mechanical property relationship in a low carbon Nb-Cu microalloyed steel processed through a three-step heat treatment: The effect of tempering process[J], Mater. Sci. Eng. A, 2015, 620: 454 - 462.

[5] 周文浩,谢振家,郭晖,尚成嘉.700 MPa 级高塑低碳低合金钢的多相组织调控及性能[J],金属学报,2015, 51(4): 407 - 416.

[6] Z.J. Xie, Y.Q. Ren, W.H. Zhou, J.R. Yang, C.J. Shang. Stability of retained austenite in multi-phase microstructure during austempering and its effect on the ductility of a low carbon steel[J], Mater. Sci. Eng. A, 2014, 603: 69 - 75.

[7] Z.J. Xie, S.F. Yuan, W.H. Zhou, J.R. Yang, H. Guo, C.J. Shang. Stabilization of retained austenite by the two-step intercritical heat treatment and its effect on the toughness of a low alloyed steel[J], Mater. Des., 2014, 59: 193 - 198.

[8] Z.J. Xie, G. Han, W.H. Zhou, C.Y. Zeng, C.J. Shang. Study of retained austenite and nano-scale

precipitation and their effects on properties of a low alloyed multi-phase steel by the two-step intercritical treatment[J], Mater. Char., 2016, 113: 60-66.

[9] 谢振家,尚成嘉,周文浩,吴彬彬.低合金多相钢中残余奥氏体对塑性和韧性的影响[J],金属学报, 2016, 52(2): 224-232.

[10] S. Liu, Z. Xiong, H. Guo, C. Shang, R.D.K. Misra. The significance of multi-step partitioning: Processing-structure-property relationship in governing high strength-high ductility combination in medium-manganese steels[J], Acta Mater., 2017, 124: 159-172.

[11] S. Liu, V. Challa, V. Natarajan, R.D.K. Misra, D. Sidorenko, M. Mulholland, M. Manohar, J. Hartmann. Significant influence of carbon and niobium on the precipitation behavior and microstructural evolution and their consequent impact on mechanical properties in microalloyed steels[J], Mater. Sci. Eng. A, 2017, 683: 70-82.

[12] S. Liu, H. Tan, H. Guo, C. Shang, R.D.K. Misra. The determining role of aluminum on copper precipitation and mechanical properties in Cu-Ni-bearing low alloy steel[J], Mater.Sci. Eng. A, 2016, 676: 510-521.

[13] G. Han, B. Hu, Y.S. Yu, X.Q. Rong, Z.J. Xie, R.D.K. Misra, X.M. Wang, C.J. Shang. Atomic-scale study on the mechanism of formation of reverted austenite and the behavior of Mo in a low carbon low alloy system[J], Mater. Char., 2020, 163: 110269.

[14] G. Han, C.J. Shang*, R.D.K. Misra, Z.J. Xie, Solid phase transition of Cu precipitates in a low carbon TRIP assisted steel[J], Physica B: Condensed Matter, 2019, 569: 68-79.

[15] G. Han, Z.J. Xie, B. Lei, W.Q. Liu, H.H. Zhu, Y. Yan, R.D.K. Misra, C.J. Shang. Simultaneous enhancement of strength and plasticity by nano B2 clusters and nano-γ phase in a low carbon low alloy steel[J], Mater. Sci. Eng. A, 2018, 730: 119-136.

[16] G. Han, Z.J. Xie, Z.Y. Li, B. Lei, C.J. Shang, R.D.K. Misra, Evolution of crystal structure of Cu precipitates in a low carbon steel[J], Mater. Des., 2017, 135: 92-101.

[17] G. Han, Z.J. Xie, L. Xiong, C.J. Shang, R.D.K. Misra, Evolution of nano-size precipitation and mechanical properties in a high strength-ductility low alloy steel through intercritical treatment [J]. Mater. Sci. Eng. A, 2017, 705: 89-97.

Study on the Applicability of Warm Forming Ultrahigh Strength Medium-Mn Steel

Chang Ying[1], Wang Cunyu[2], Li Xiaodong[1], Zheng Guojun[1], Dong Han[2,3]

1 School of Automotive Engineering, Dalian University of Technology, Dalian 116024, China   2 Central Iron & Steel Research Institute (CISRI), Beijing 100081, China   3 School of Materials Science and Engineering, Shanghai University, Shanghai 200444, China

The hot-formed ultrahigh-strength steel holds a tensile strength of 1 500-

2 200 MPa and its application has been an index to passive safety and lightweight of an automobile. However, the available stamping process of hot-formed steels in production has a narrow process window, in which the heating temperature above 900 ℃ and the cooling rate above 30 ℃/s are necessary. The high cooling rate makes the heat transfer efficiency of the steel plate different, which results in the martensitic transformation difficult to complete all over the part. Consequently, the uneven mechanical properties are caused in different areas, such as the side wall and the bottom of the part. Moreover, the steel plate surface needs Al-Si coating protection in production, but this technology is monopolized by Arcelor-Mittal (AM). Based on the existing problems, our research team developed a kind of 1 500 - 2 200 MPa ultrahigh-strength medium-Mn automobile steel and its supporting technology. The proposed medium-Mn steel has simple composition (0.1C5Mn), low heating temperature and cooling rate (wider industrial window), low cost (no Al-Si coating required), better performance and rigidity, as well as uniform thickness of parts. Its applicability technology research has also been carried out. Compared with other hot-formed steels, the medium-Mn steel has the following advantages: No decarburization layer; Uniform distribution of properties and thickness; High plasticity and toughness under the same strength level, etc.

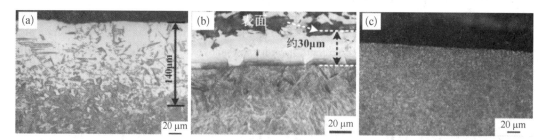

Fig. 1 (a) uncoated 22MnB5 steel, (b) Al-Si coated 22MnB5 steel and (c) uncoated medium-Mn steel

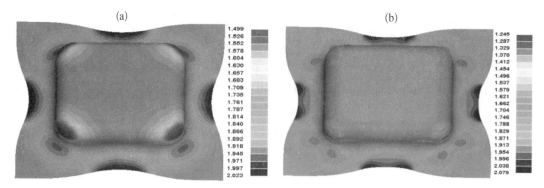

Fig. 2 Thickness distribution diagram of deep drawing part of (a) medium-Mn steel (between 1.499 and 2.023 mm) and (b) 22MnB5 steel (between 1.245 and 2.079 mm)

Strengthening and Toughening Mechanism for High-alloyed Martensitic Steel

Luo Haiwen[1,*], Wang Xiaohui[1,2*], Liu Zhenbao[2], Yang Zhiyong[2]

1 Department of Ferrous Metallurgy, University of Science and Technology Beijing, 30 Xue Yuan Lu, Beijing, China

2 Division of Special Steels, Central Iron and Steel Research Institute, 76 Xue Yuan Nan Lu, Beijing, China

【Introduction】

Martensite, a typical hierarchical microstructure in the quenched low-alloyed steels, can be characterized at multiscale, including martensite packet, block and lath at the descending size. What microstructural unit determines the strength and toughness of Ultra-high strength stainless steel (UHSSS) containing high contents of Cr, Ni, Co and Mo is then of great value on developing UHSSS with improved mechanical properties for the land gear of aircraft. In this work, the various hierarchical martensitic microstructures were manufactured by employing different manufacturing processes and then characterized at multiscale, their correlations to strength and toughness were then analyzed.

【Experimental】

A UHSSS having the composition of 0.2C-12.4Cr-13.8Co-5.0Mo-2.1Ni-(W+V) was melt and cast to an ingot before hot forging into several 60 mm-thick billets. They were then homogenized at 1 200 ℃ for 1 h, and hot rolled to either 15 mm with the finish temperature at 1 050 ℃ (75% reduction in thickness, HR75) or to 30 mm (50% reduction) at both 1 050 ℃ (HR50) and 1 080 ℃ (HR50 - 2), followed by quenching in oil. Some HR75 specimens were reheated to 1 080 ℃ for 1 hour and quenched in oil (HR75 - 2). The mechanical properties and microstructures were then examined for all of them.

【Results】

The sizes of prior austenite grains and martensite packets/blocks/laths are listed in

Table 1 together with the measured impact toughness and yield strength. The relationship between YS and all the microstructural unit sizes are shown in Fig.1 and then fitted by the Hall-Petch equation. It is found that the relation of "effective grain size", defined by high angle boundaries (HABs), and YS can be best fitted by the Hall-Petch, suggesting that it may be the key microstructural parameter. This is reasonable since strengthening results from the resistance to dislocation gliding across grain boundaries, and HABs have greater resistance than LABs because dislocations cannot cross HABs but pile up near them.

Table 1　The hierarchical martensitic microstructural features, including the average sizes of prior austenite grains (d_a), martensite packets (d_p)/blocks (d_b)/laths (d_l) and the aspect ratio of martensite lath, and mechanical properties, including both Charpy U-Notch impact toughness (CUN) and yield strength ($0.2\% \ offset \ proof \ stress, \ YS$), were all measured on UHSSS specimens subjected to four different manufacturing processes

Sample No.	Hot rolling *Process*	d_a/μm	d_p/μm	d_b/μm	d_l/μm	Aspect ratio	CUN/J	YS/MPa
HR75	75% reduction and finished at 1 050 ℃	6.3±3.7	2.8±1.1	2.1±1.0	0.24±0.11	11.6	186±3	1 065±4
HR50	50% reduction and finished at 1 050 ℃	14.5±8.8	6.3±2.7	2.1±0.9	0.26±0.12	24.2	193±3	1 035±2
HR50-2	50% reduction and finished at 1 080 ℃	16.1±9.1	6.7±2.2	2.2±0.9	0.25±0.11	21.9	190±5	1 020±2
HR75-2	HR75 reheated to 1 080 ℃ for 1 h	76.1±45.3	24.0±10.6	2.7±1.0	0.44±0.18	56.6	125±1	913±3

Fig. 2 shows these voids and cracks in the fractured specimens after the impact test and the nearby microstructures in the matrix including no cracks. It appears that cracks propagate across martensite laths by shearing them and along the block boundaries, as indicated in red and blue lines respectively. When a crack is propagating across martensite lath and encounters an austenitic film at the lath boundary, it may either continue into austenite or be deflected to a different path. The former may cause strain-induced transformation in austenite, while the latter leads to much longer crack, both improve toughness. The film-like RA with a width of about 10 nm at lath boundaries was indeed observed under TEM, see Fig. 2(c). The thinner laths mean that the propagating cracks should encounter more RA films at lath boundaries, i.e. the stronger hindrance to cracking. In addition, some coarse RA grains could also resist cracking, but their resistance should be weaker since their quantity and mechanical stability are all lower than those at the lath boundary.

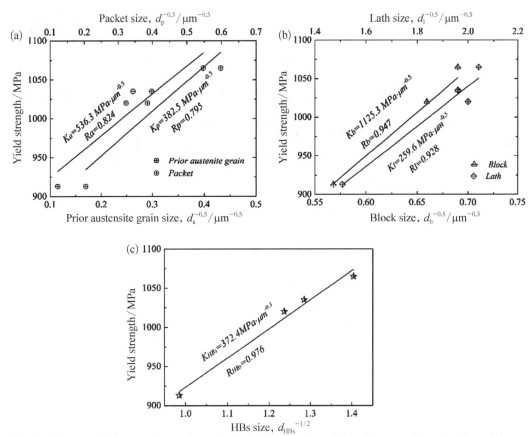

Fig. 1 Plots of yield strength *vs.* the reciprocal square root of (a) prior austenite grain size (d_a) and packet size (d_p), (b) block width (d_b) and lath width (d_l) as well as (c) the size defined by high angle boundaries (d_{HBs}). All of them are fitted by Hall-Petch equation with the correlation coefficient, R

The following conclusions may be drawn:

(1) When the coarse prior austenite grains in UHSSS were refined to about 16 μm, all the multi-level martensitic units, including martensite packet/block/lath, were also clearly refined correspondingly, leading to the substantial increases in both YS and toughness. In contrast, the further refinement to about 6 μm led to the decrease in packet size but negligible changes in both block and lath sizes and the resultant yield strength increased marginally and impact toughness changed negligibly.

(2) It is concluded that the increase of yield strength follows the Hall-Petch dependence on the effective grain size, which is defined by high angle boundaries (HAB) including the prior austenite grain boundaries, the martensite packet, and block boundaries in UHSSS.

(3) The impact toughness of UHSSS is actually determined by the quantity, the size, the distribution, and mechanical stability of austenite grains retained at both HABs and lath boundaries because the propagating cracks can be trapped by them,

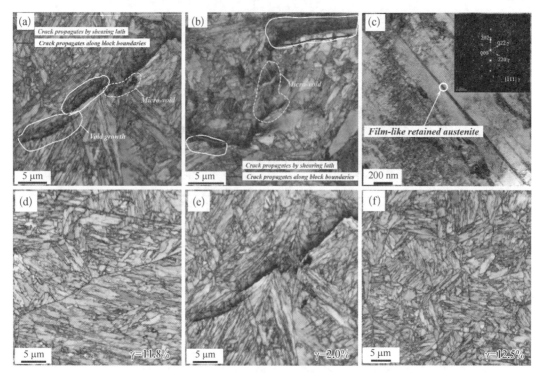

Fig. 2 EBSD misorientation and boundary image mapping on the propagating cracks in HR50 (a) and HR75 (b) after impact fracture; Bright-field TEM image and SAED of film-like retained austenite in HR50 (c); EBSD image quality and phase mapping on the matrix of HR50 without cracks (d), near cracks (e) and on HR75 (f). The green represents FCC phase.

leading to improved toughness. In particular, the fine film-like RA grains at lath boundaries may contribute more to toughness than those at HABs because both their number density and mechanical stability are higher. When the refinement of hierarchical martensite microstructure leads to the significant increase in the density of both HABs and lath boundaries, this may produce a higher number of RA grains having the finer size and enhanced mechanical stability; thus improving toughness indirectly.

More details can be found at doi: 10.1016/j.jmst.2020.04.001

Physical Metallurgy-guided Machine Learning and Artificial Intelligent Design of Ultrahigh-strength Stainless Steel

Shen Chunguang, Wei Xiaolu, Wang Chenchong, Xu Wei*

State Key Laboratory of Rolling and Automation, Northeastern University

* Corresponding author: xuwei@ral.neu.edu.cn

【Introduction】

With the development of data mining, machine learning (ML) has been widely applied in the discovery of new material with improved properties. However, ML belongs to a statistical method that directly establishes the relation between input parameters and target properties, not involving the physical metallurgy (PM), which limits the design process into a mathematical process and ignores the important role of PM. Thus, such a design process may mislead the discovery of new materials and reduce design efficiency. To address this challenge, the present work firstly compares the strength of PM and PM methods in property prediction and alloy design using various datasets of ultra-high-strength (UHS) steel. The results show that has a strong ability to explore alloy space but some design solutions violating PM. The performance of the PM method is sensitive to the alloy systems, so it is difficult to be applied for large-scale alloy space. Based on the above conclusions, we propose a novel PM-guided ML method, in which intermediate PM parameters are introduced into the original dataset to participate in and guide the ML process, and a design framework is developed combining PM-guided ML regression, ML classifier, and genetic algorithm (GA), and a new UHS steel is successfully designed and validated. Moreover, compared to alloy design using pure ML, the present framework can improve design efficiency and produce solutions consistent with PM principles.

【Experimental】

Several datasets were extracted from literature, in which compositions and aging conditions were used as inputs and hardness were used as output. With respect to ML, SVM and ANN were selected to establish the correlation between composition/aging condition and hardness. For the PM method, two widely accepted PM models were applied to describe the whole PM process and predict hardness. Then, alloys with higher hardness are designed combining predicted models and GA. Finally, the advantages and weaknesses of the two methods are compared based on prediction and design results.

The PM-guided ML framework can be divided into the following parts: (1) dataset construction collecting from literature; (2) PM-guided prediction model is developed using inputs combining with PM parameters; (3) alloy design is performed combining the PM-guided ML model and GA to explore alloy space; (4) establishing an intelligent screening method to determine the promising alloys and experiment alvalidation.

【Results】

Through a systematic comparison between machine learning and physical metallurgy model, it is found that: with respect to performance prediction, the ML method has stronger prediction ability. The comparison results are shown in Fig.1, while the PM model has limited prediction accuracy due to the limitation of alloy sensitivity. Compared with the ML method, the PM model has lower data dependence and stronger explainability of a physical mechanism. For alloy design, the ML method can fully explore the alloy space, while the PM model can obtain the results following the principles of physical metallurgy, so the combination of the two advantages can be well applied in the material discovery.

For alloy design based on PM-guided ML, 500 SVM-PM model was used as the objective function of GA, and GA was applied to find optimal alloys within the dataset

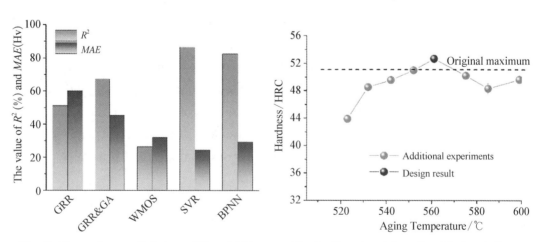

Fig. 1 Comparison results between ML and PM   Fig. 2 Experimental results of designed alloy

range considering the present small sample. After screening 500 SVM-PM model results, 116 optimal solutions were finally determined. Subsequently, the SVC classifier was established to further screen the high-reliability solutions from the 116 results mentioned above. Finally, two alloys were designed as shown in Table 1, and alloy 2 was selected for experimental verification. The experimental results show that the designed hardness is in good agreement with experimental results.

Effect of Surface Deformation on Stress Corrosion Crack Initiation in Austenitic Stainless Steels in PWR Primary Water

Chang Litao*, M. Grace Burke, Fabio Scenini

Materials Performance Center, The University of Manchester, Manchester, UK M13 9PL

* Corresponding author: litao.chang@manchester.ac.uk

【Introduction】

Austenitic stainless steels are widely used in the nuclear power plants due to their good general corrosion resistance to the high temperature aqueous environment. However, they can suffer from environmentally-assisted degradation problems, such as stress corrosion cracking (SCC), during the long-term exposure to the environment. Numerous researches indicate that cold-work, induced either intentionally or incidentally, is necessary for SCC in austenitic stainless steels in PWR primary water. Machining is an important step in manufacturing, machining always introduces deformation in the surface region of the material. In the present study, the effect of the machining on SCC initiation of austenitic stainless steels in PWR primary water has been investigated.

【Experimental】

Warm-forged stainless steel was used in the present research, and the machined surface was prepared through surface milling. Flat tensile specimens were extracted from the machined block for the tests, and the other side of the specimen was polished to a mirror surface finish. A Slow strain rate tensile test was used to accelerate the crack initiation in high-temperature hydrogenated water, and the strain rate used was 2×10^{-8} s^{-1}. The Microstructure of the forged substrate, surface deformation layer, and the cracking behaviors were characterized using SEM and TEM.

【Results】

The machining-introduced deformation layer is characterized by an ultrafine-grained outer layer and a deformed inner layer [Fig. 1(b)]. SSRT test results showed that machining significantly reduced the SCC initiation susceptibility of this stainless steel as a reduced number of cracks were identified in the machined surface [Fig. 1(d)] compared to the polished surface [Fig. 1(c)]. This beneficial effect of machining was attributed to the huge amount of grain boundaries within the ultrafine-grained layer on the machined surface which promoted more uniform oxidation, and it is suggested that the microstructural features may play a more important role than stress during SCC initiation. The associated mechanisms and

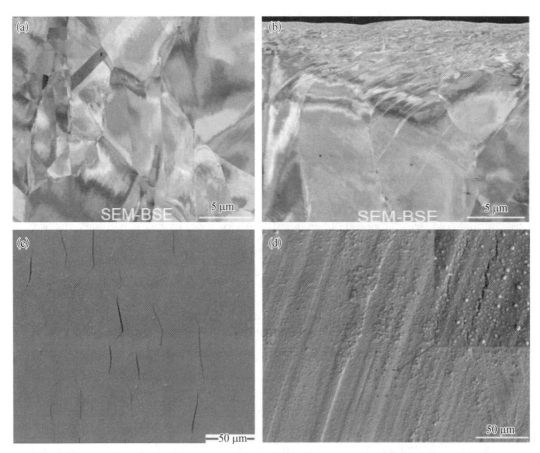

Fig. 1 cross-sectional microstructure of polished surface (a) and machined surface (b) of the material; crack initiation on polished surface (c) and machine surface (d)

possible implications of the results have been discussed.

刀具用高碳马氏体不锈钢碳化物调控研究

赵洪山*,杨玉丹,彭伟,董瀚

上海大学

*通讯作者:boyushankf@126.com

【引言】

我国五金刀剪行业面临着转型升级的战略需求,高端刀具用钢市场容量及潜在需求量大。而国内高端刀具用钢主要依赖进口。高端刀具用马氏体不锈钢主要是碳含量0.4%以上的高碳马氏体不锈钢,除高硬度需求外,高端刀具用钢还对韧性与耐蚀性要求较高。这主要与冶金质量与碳化物调控技术有关。随着冶炼设备与工艺的改进,不锈钢纯净度的提高,高碳马氏体不锈钢中碳化物的含量、分布及尺寸大小逐步成为决定其性能的关键因素。

【方法】

以多相、亚稳、多尺度M3组织调控技术与多形合金化概念为指导,通过初炼、精炼、电渣重熔、锻造、热轧、热处理全流程控制调控马氏体不锈钢中碳化物,进而提高其性能与品质。主要包括成分优化,成分设计上避开包晶区域,以减少一次碳化物析出量。严格控制成分准确度,降低杂质元素的含量。通过特定的稀土与Ag添加技术,获得高耐蚀与抗菌钢板。通过优化电渣重熔、锻造与热轧参数,降低进而消除一次碳化物的危害。在随后球化退火及热处理过程中,优化调控二次碳化物的含量、分布及尺寸大小。

【结果】

通过以上调控方法,基本可消除6Cr16MoMA高碳马氏体不锈钢中的一次碳化物,且退火态钢板二次碳化物尺寸较小,分布均匀,最大二次碳化物≤2 μm,平均二次碳化物尺寸≤0.8 μm,如图1(a)、(c)所示。产品1 025 ℃~1 100 ℃奥氏体化后淬火,碳化物形貌、分布、尺寸大小分别如图1(b)、(c)所示。随着奥氏体化温度增加,二次碳化物含量及平均尺寸随之降低,未出现二次碳化物平均尺寸升高(1 025 ℃~1 100 ℃分别为0.47 μm、0.44 μm、0.45 μm、0.41 μm),再次表明二次碳化物尺寸均匀,无异常大颗粒,表明上述高碳马氏体不锈钢碳化物调控手段有效。

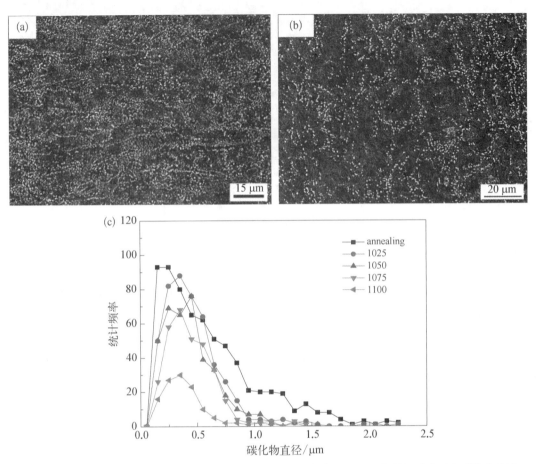

图 1 6Cr16MoMA 马氏体不锈钢微观组织及统计结果 (a) 退火态; (b) 1 025 ℃淬回火态; (c) 碳化物统计

6Cr16MoMA 热处理后硬度可达 56～59 HRC, 120 ℃～200 ℃回火后, 硬度基本不变, 甚至略有提高, 抗拉强度最高约 2 200 MPa。与国内同厂商用高碳马氏体不锈钢相比(5Cr15MoV、7Cr17MoV、9Cr14MoV、9Cr18MoV), 韧性翻番, 高硬度下半标试样(55 mm×10 mm×5 mm) KU_2 可达 8～11 J; 抗点蚀、盐雾能力、耐磨性明显更优, 并可以有效抵抗大肠杆菌、金黄色葡萄球菌、铜绿假单胞菌等细菌, 抗菌率≥99.99%, 如表 1 所示。该材料制作的厨刀性能优异, 锋利度、耐用度、耐蚀性高, 且抗菌效果明显。

表 1 6Cr616MoMA 钢抗菌试验结果, JISZ2801 标准

试验菌种	对照样 0 h 菌落数 $A/(CFU·mL^{-1})$	对照样 24 h 菌落数 $B/(CFU·mL^{-1})$	试样 24 h 菌落数 $C/(CFU·mL^{-1})$	抗菌活性 /R	抗菌率 /%
大肠杆菌	$2.5×10^5$	$2.9×10^6$	62	4.67	>99.99
金黄色葡萄球菌	$2.4×10^5$	$3.5×10^5$	<10	>4.54	>99.99
铜绿假单胞菌	$1.7×10^5$	$5.7×10^5$	<10	>5.76	>99.99

High Nitrogen Steels: Livelihood Applications

Peng Wei[1,3,*], Gao Xinqiang[1], Zhao Hongshan[1,3], Weng Jianyin[2], Liu Tengshi[1,3], Xu Dexiang[1,3], Li Bei[2], Dong Han[1,3,*]

1 Shanghai University 2 FIYTA (Group) Holdings Co. Ltd.
3 Shanghai University New Materials (Taizhou) Research Institute
* Corresponding author: pw20080607@126.com, 13910077790@163.com

【Introduction】

Health is the most important issue for people concern. Livelihood metallic products, such as watches, belt buckles, biomaterials, accessories, are mainly made of stainless steel. Traditional stainless steel contains 8 – 10 wt.% nickel, its products will induce skin allergy and cytotoxicity due to nickel release when long-term use in people's body. Thereby, nickel content needs to be restricted in the human body. Low nickel or nickel-free high nitrogen steels (HNS) with an excellent combination of strength and ductility, good corrosion resistance, good biocompatibility, anti-allergy and non-magnetic, have become prospective steel for manufacturing livelihood products and applications.

【Experimental】

Industrial hot-rolled (10Cr21Mn17N) and cold-rolled (05Cr21Mn16Ni2N) HNS plates were fabricated with nitrogen content up to 0.6 wt.%. Mechanical properties, artificial sweat corrosion tests, nickel releases, and product manufacture were conducted to investigate HNS. SEM, OM, EBSD were used to explore the microstructure and properties.

【Results】

As known to all, nitrogen can stable and strengthen the austenitic matrix. Both the hot-rolled 10Cr21Mn17N [Fig.1(a)] and cold-rolled 05Cr21Mn16Ni2N plates are fully austenitic microstructure after solution treatment. Furthermore, the strength can be largely improved by cold deformation. The yield strength immediately increases

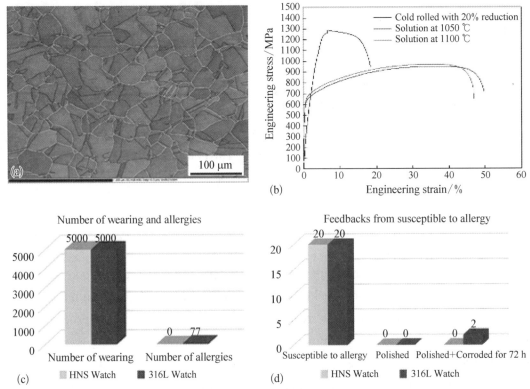

Fig. 1 (a) EBSD microstructure of 10Cr21Mn17N, (b) Tension results of 05Cr21Mn16Ni2N, (c) Number of wearing and allergies, (d) Feedbacks from susceptible to allergy

from ~570 MPa to ~1 160 MPa and the tensile strength raises by 330 MPa after suffering a 20% cold-rolled reduction. For all that, the cold-rolled reduction sample remains still 22% elongation. In addition, solution treatment at 1 050 ℃ and 1 100 ℃ temperature has little influence on the strength and ductility of 05Cr21Mn16Ni2N steel, as shown in Fig.1(b)[1].

Livelihood products, such as watches and belt buckles, were made of high nitrogen steels. In Fig.1(c-d), the number of wearing HNS and 316L watches, and their allergies indicated that people were more susceptible to allergy after wearing 316L watches, especially suffering sweat corrosion. Nevertheless, the group wore HNS watches that presented no allergies. Standard EN1811: 2011 + A1: 2015 limited nickel release to 0.2 $\mu g/(cm^2 \cdot week)$ for piercing and 0.5 $\mu g/(cm^2 \cdot week)$ for non-piercing, respectively. The tested nickel release of 05Cr21Mn16Ni2N steel is less than 0.1 $\mu g/(cm^2 \cdot week)$. Moreover, the simulated artificial sweat test indicated that HNS livelihood products are more resistant to corrosion than 316L and 304 stainless plates of steel. Hence, HNS is suitable for a wide of livelihood steel materials.

Fig. 2　HNS livelihood products: Anti-allergic watches and belt buckles, non-magnetic phone lens decoration circle

【Acknowledgement】

This work was financially supported by Project funded by the China Postdoctoral Science Foundation (2019M651465).

References:

[1] Dong H, et al. High Performance Steels: The Scenario of Theoryand Technology[J]. Acta Metall Sin, 2020, 56: 558-582.

Investigations on the Stress Corrosion Cracking Initiation Behavior of 316LN Stainless Steel in PWR Primary Water

Zhong Xiangyu*, Tetsuo Shoji

Tohoku University

*Corresponding author: zhong.xiangyu.d4@tohoku.ac.jp (Xiangyu Zhong)

【Introduction】

Austenitic stainless steels have been widely used in nuclear power plants because of their good corrosion resistance in the high-temperature water. However, environmentally assisted degradation problems such as stress corrosion cracking (SCC) during prolonged exposure to the high temperature water have been observed in the Austenitite stainless steel components in the nuclear power plants. Several cases of SCC have been identified in the non-sensitized austenitic stainless steel components of several boiling water reactors (BWR) and pressurized water reactors (PWR) over the past years. The effects of several variables, such as water chemistry, testing temperature, degree, and paths of the pre-strain have focused on understanding the phenomenology of SCC of stainless steels in the nuclear power plant. However, most of these studies have concentrated on the crack propagation stage rather than initiation, the mechanistic understanding of the SCC initiation is still missing.

【Experimental】

The material used for this study was a solution-annealed 316LN with the composition of (in wt.%): C: 0.01, Si: 0.48, P: 0.023, Cr: 17.5, Mn: 1.42, Ni: 11.64, Mo: 2.08, N: 0.116, O: 0.047, S < 0.003, the balance being Fe. The geometry, dimension, and preparation processing of the hollowed cylindrical SSRT specimen was detailed described in our previous work and described briefly in the following texts. The surface finish of the inner surfaceof the specimens was formed by two different machining processes. One is by drilling (marked as 316LN-drilled) where a drill with 4 mm in diameter was used. The other surface finish was honed after drilling (marked as 316LN-honed) to remove the heavily machined surface and also to make the surface smoother.

The test was performed using an SSRT system with a circular high-temperature water loop system. The test environment is simulated PWR primary water with 500 ppm B (as H_3BO_3), 2 ppm Li (as LiOH), dissolved oxygen (DO) < 10 ppb at 325 ℃, 15.6 MPa. DH in the inlet water was controlled by bubbling either pure nitrogen gas, pure hydrogen gas, or a mixture of nitrogen and hydrogen gases in the reservoir water tank until equilibration reached. The DH levels (15 cm^3 (STP) H_2/kg H_2O) were used in this study. The strain rate of the SSRT test in the present study is about $2 \times 10^{-7} s^{-1}$. The tests were interrupted when the load reached the maximum on the Load-Displacement curves (marked as 316LN-drilled-max, 316LN-honed-max) and the strain of 7.5% (marked as 316LN-drilled-7.5%, 316LN-honed-7.5%). After an SSRT test, a longitudinal cross-section was obtained by electrical discharge wire-cutting and metallographic polishing and the depth of the cracks on the cross-section was measured by optical microscope (OM) and scanning electron microscope (SEM).

【Results】

Fig.1 is the stress-strain curves of 316LN different surface conditions in the PWR

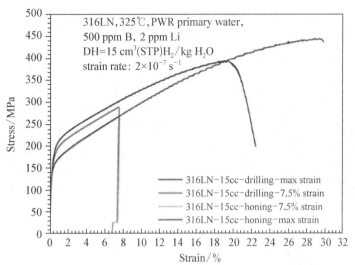

Fig. 1 The Stress-strain curves of 316LN in PWR environment

environment with DH level = 1 515 cm³ (STP) H_2/kg H_2O, which were calculated from Load-Displacement curves. The strain-stress curves show that the elongations of the specimens were affected by the surface condition significantly. The elongations of the honed specimens are larger than the drilled specimen. It means that the SCC susceptibility of a honed specimen is lower than the drilled specimen.

Fig. 2 shows the optical morphologies of the inner surface of 316LN-drilled and 316LN-honed in the PWR environment with DH=15 cm³ (STP) H_2/kg H_2O. Many cracks were observed along the gauge length direction in all specimens irrespective of surface finish condition. Even though in the specimens interrupted at the strain of 7.5%, lots of tiny cracks were observed on the inner surface along the gauge length direction. It indicates that the crack initiated at the early stage before the maximum load. The continued nucleation of new cracks may result in the reactivation of dormant cracks and coalescence between the dormant crack and a subsequently nucleated crack. The number and length of cracks for the drilled and honed specimen are statistically analyzed based on the optical images. The analyzed area is about 11.44 mm². The cracks in the 316LN-drilled specimen are longer than that of the 316LN-honed specimens. The histograms of crack traces distribution of 316LN after SSRT in the PWR environment was shown in Fig.3a. The statistical results show that the length of most cracks less than 0.2 mm for both drilled and honed specimen. The number of the crack in the 316LN-drilled specimen isless than that of the 316LN-honed specimen (as shown in Fig.3b). The number of cracks for the specimen interrupts at the 7.5% strain is much larger than that of an interrupt at the maximum strain for both drilled and honed specimen. The number of crack decreases may relate to the crack

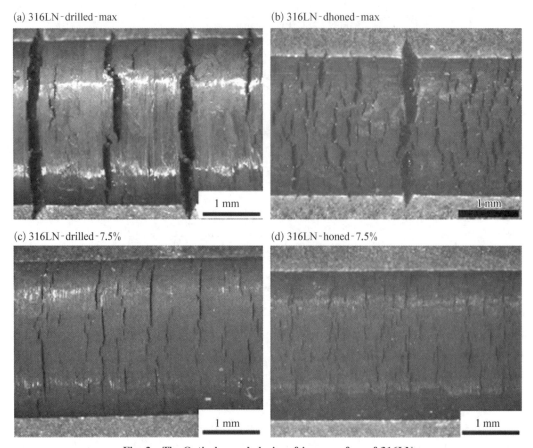

Fig. 2　The Optical morphologies of inner surface of 316LN

coalescence during the SSRT test. Fig.4 shows the EBSD analysis results of the crack tip. Both transgranular crack (TG crack) and intergranular crack (IG crack) were observed. It is also found that the crack initiates as a TG crack, and then transform to an IG crack during the crack propagation.

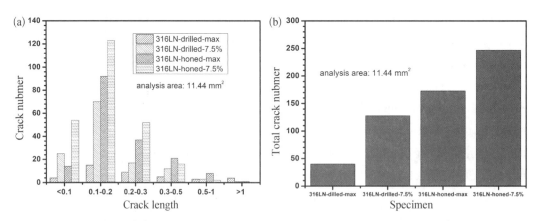

Fig. 3　(a) Histograms of crack traces distribution of 316LN after SSRT;
(b) Total crack number of 316LN after SSRT in PWR environment

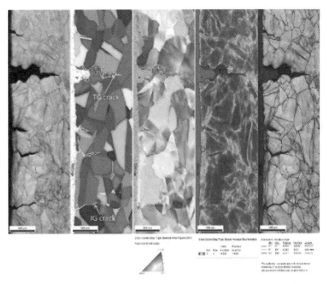

Fig. 4 EBSD results of typical crack tip of 316LN after SSRT

High Energy Storage of a Metallic Glass via Long-term Cryogenic Thermal Cycling

Bian Xilei[1,2,*], Wang Gang[2,*], Ding Ding[2], Christoph Gammer[1], Gerhard Wilde[3], Jürgen Eckert[1,4]

1 Austrian Academy of Sciences 2 Shanghai University
3 University of Münster 4 Montanuniversität Leoben
* Corresponding authors: g.wang@shu.edu.cn; Xilei.Bian@oeaw.ac.at

【Introduction】

Due to the nature of the kinetic metastability, the quenched metallic glasses (MGs) with a relatively high energy state tend to transform to a lower energy state via atomic rearrangement, which can largely reduce the amount of free volume and further deteriorate their mechanical properties such as ductility. Therefore, it is of crucial importance to modulate the energy states to attain the desired ductility of MGs. In this work, we choose a Zr-based MG as a model material for long-term cryogenic cycling. Heat capacity measurement implies that long-term thermal cycling can drive the MGs to a higher energy state, implying that long-term thermal cycling is an effective method for rejuvenation of MGs. By adjusting the long-term thermal cycling time, we demonstrate that time has strong effects on the degree of structural rejuvenation and mechanical properties. The underlying mechanism for rejuvenation and the change of mechanical properties is discussed.

【Experimental】

Thermal cycling was performed for a different time ranging from 0 to 55 days. Compression tests were conducted using an MTS CMT5205 machine with a strain rate of $2.5 \times 10^{-4} s^{-1}$ at room temperature. Heat capacity C_p were measured by a Quantum Design physical properties measurement system (PPMS 6 000) under high vacuum conditions in the temperature range of 77 - 302 K. Nanoindentation tests were carried out to obtain the hardness and elastic modulus.

【Results】

As the thermal cycling time (t_{CT}) increases, the value of C_p first increases from 0 to 25 days and then decreases. It should be pointed out that the value of C_p is higher than that of the as-cast sample even after 50-day CT. This implies that thermal cycling can effectively render an enhancement of C_p and the degreeof this enhancement is strongly dependent on process time (Fig.1a). Compared with the as-cast sample, the strength is significantly decreased and the plasticity is improved after thermal cycling. Fig. 1(b) clearly shows that the yield strength displays a reverse change trend with respect to the entropy.

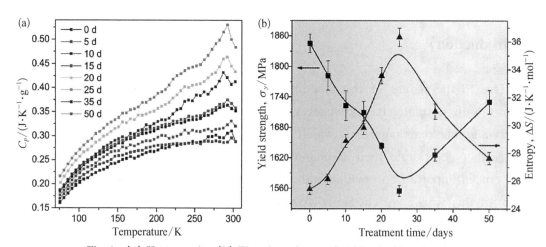

Fig. 1 (a) Heat capacity. (b) Time dependences of yield strength and entropy

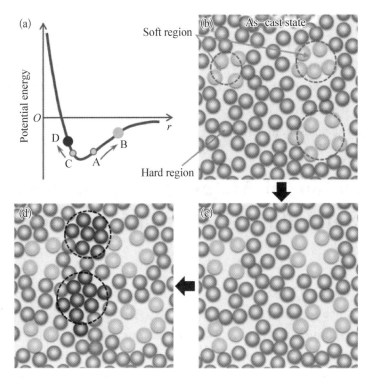

Fig. 2 Schematic representation of the atomic mechanism for the rejuvenation

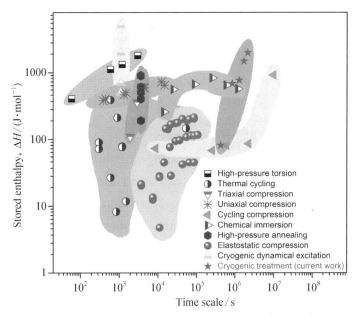

Fig. 3 Comparison of the stored enthalpy by various methods

Since the activation volume of an STZ is directly related to the distribution hardness, smaller STZs are expected to emerge in glasses with pronounced structural heterogeneity, and vice versa. Nanoindentation tests reveal that the corresponding mechanical response is a consequence of the convolution of the intrinsic thermal activation process and the structural heterogeneity in the glassy phase. Due to the introduction of thermal cycling-induced structural inhomogeneities, the distribution of energy barriers for shear transformations becomes broader, and the lowest activation barrier value is decreased. Thus, the mechanical performance of MGs can be tuned by controlling the thermal cycling time.

Fabrication of Metal Matrix Composites by Solid-state Cold Spraying Process

Xie Xinliang[1,2,*], Ji Gang[1,2], Chen Chaoyue[3], Chen Zhe[4], Liao Hanlin[1]

1 ICB UMR 6303, CNRS, Univ. Bourgogne Franche-Comté, UTBM, F-90010 Belfort, France 2 Univ. Lille, CNRS, INRAE, Centrale Lille, UMR 8207-UMET-Unité Matériaux et Transformations, F-59000 Lille, France
3 State Key Laboratory of Advanced Special Steels, School of Materials Science and Engineering, Shanghai University, Shanghai 200444, China
4 State Key Laboratory of Metal Matrix Composites, Shanghai Jiao Tong University, Shanghai 200240, China
* Corresponding author: xinliang.xie@utbm.fr

【Introduction】

As a solid-state powder deposition process, the emerging cold spray additive manufacturing (CSAM) is attracting increasing attention in recent years. Due to its unique 'cold' feature, the CS technique can avoid the serious grain growth, strong columnar grains, and phase change in comparison with the fusion-based AM techniques by laser, electron beam, and/or arc. Recently, various efforts have been made in the study of metal matrix composites (MMCs) by CSAM, demonstrating great application potential in aerospace, automotive, and electronics industries This article aims to investigate the CS-processed MMCs from the aspects of composite powder design, deposition processing, microstructure evolution, mechanical and corrosion properties. Furthermore, post-treatments including heat treatment and friction stir processing were applied to further improve the mechanical properties of cold-sprayed composites by reducing the defects, modifying the microstructure, and more importantly enhancing the interface bonding. Finally, the challenges and future perspectives on the fabrication of advanced MMCs by CS are addressed.

【Experimental】

A specially designed composite powder reinforced with nano/submicron-TiB_2 particles (7.0 wt. % TiB_2/AlSi10Mg) was produced by in-situ chemical reaction of the mixed salts in molten Al followed by gas atomization. Both the gas-atomized pure AlSi10Mg and TiB_2/AlSi10Mg composite powder were used as the feedstocks and they were deposited onto grit-blasted Al substrates (140 mm×80 mm×4 mm) by using a homemade helium circulation CS system (LERMPS, UTBM, France). Helium was used as the propellant and carrier gas with an inlet pressure and temperature of

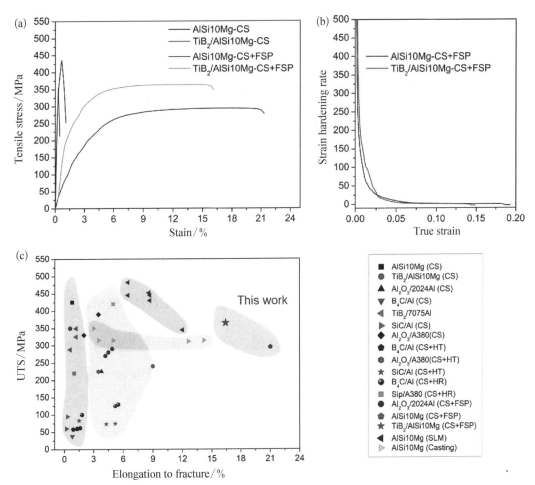

Fig. 1 (a) Tensile stress-strain curves for pure AlSi10Mg and TiB_2/AlSi10Mg composites before and after FSP treatment. (c) Comparison of the UTS and elongation values of the FSP treated pure AlSi10Mg and TiB_2/AlSi10Mg composite deposits with other AMCs processed by CS, CS+heat treatment (HT), CS+hot rolling (HR), and CS+FSP treatments

1.8 MPa and 320 ℃, respectively. The nozzle standoff distance and traverse speed were set 30 mm and 100 mm/s, respectively. The nozzle trajectory was repeated for more than 30 times to obtain thick deposits (>2 mm). Then, FSP treatments were performed using a commercial friction stir welding (FSW) machine (FSW-RL31-010, Beijing FSW Technology Co., Ltd, China). An H13 steel stir-tool with a threaded pin of 3.5 mm in root diameter and 2.0 mm in length, and a concave shoulder of 10 mm in diameter, was installed in this FSW equipment. The rotation speed and traverse speed were set at 1 500 rpm and 500 mm/min, respectively.

【Results】

A simultaneous enhancement in both strength and ductility of the cold sprayed TiB_2/AlSi10Mg composite deposits has been successfully achieved by a novel strategy combining gas-atomization (involving in-situ reaction), CS, and post-FSP. Alternating of the inter-splat bonding conditions from prominent interlocking or partially metallurgical bonding in the as-spray state to a fully metallurgical bonding via FSP treatment is essential for achieving a high mechanical performance of the cold-sprayed samples. The strengthened strength and ductility can be attributed to the homogenous distribution of reinforcements, matrix grain refinement, and more importantly, the robust interfacial bonding of the in-situ TiB_2 particles and Al matrix. This study provides new guidance for hybrid AM of MMCs with high performance.

One-step Sintering Synthesis of Superfine $L1_0$-FePt Nanoparticle by Using Liquid-assisted

Zhao Dong[1], Pei Wenli[1,*], Wu Chun[2], Wang Jianjun[1], Wang Qiang[1]

1 Northeastern University　2 Liaoning Technical University

* Corresponding author: peiwl@atm.neu.edu.cn

【Introduction】

FePt nanoparticles (NPs) with fct ($L1_0$) structures have been paid considerable attention for their potential applications. The chemical synthesis has become an extensive route for the fabrication of FePt NPs. Normally, the FePt NPs directly synthesized by the chemical methods is fcc structure. In order to obtain good performance, the particles should be transferred from fcc to fct structure by a post-annealing. However, the particles would grow up and aggregate during this treatment, which seriously worsens the performance of the NPs and limits their application. To solve this bottleneck problem, many efforts have been made. SiO_2, MgO, and NaCl are employed as the protective inert layer to restrict grain growth and aggregation. However, the effects are limited and abnormal large particles and aggregation could still not avoid. However, up to now, the FePt NPs prepared by direct wet chemical synthesis route show a partial structural ordering, which isn't high enough for further applications. In this work, we reported a one-step sintering method, in which a liquid-assistant process was employed. The relationships among the synthesizing process, particle size, and ordering degree were discussed. The optimal process to prepare FePt NPs with a superfine size and good performance was studied. Such a liquid-assisted method offers a good solution to fix the uniformity problem of solid-solid mixing in the previous report and is a promising facile method to synthesize superfine $L1_0$-FePt NPs.

【Experimental】

The metal precursors 0.25 mmol $Fe(acac)_3$ and 0.25 mmol $Pt(acac)_2$ were dissolved in 50 mL of hexane, and then NaCl (the mass ratio of precursors to insulator was set as 1/400 to 1/100) was added into the solution. The mixture was heated to 80 ℃ under strongly stirring. After evaporation of the solvent, the mixture was heated in a tube furnace at a rate of 10 ℃/min under the reducing atmosphere ($Ar + 5\%H_2$). Finally, the $L1_0$-FePt NPs were collected through centrifugal washing repeatedly with water and alcohol and stored in alcohol at −20 ℃.

【Results】

The $L1_0$-FePt NPs with superfine size (about 7 nm, TEM) and high coercivity (up to 2.29 T) were successfully synthesized by the liquid-assisted sintering method. With the assisting of liquid, the precursors formed a shell with uniform composition and size on NaCl mediums, which increased the nucleation rates, promoted the orderly transformation, and thus improved the hard-magnetic properties. In this process, the Fe and Pt atoms increased to equal ratio, and the *fcc-fct* transformation occurred at 600 ℃. Raising the temperature would increase the grain sizes, widen size distribution, enhance the order degree and improve the hard-magnetic properties of $L1_0$-FePt NPs. A reasonable high concentration of the precursors could also increase the grain sizes, enhance the order degree, and improve the hard-magnetic properties. Particularly, it guaranteed the uniformity of sizes. This result indicated that the *liquid*-assisted is a feasible strategy to synthesize $L1_0$-FePt nanoparticle with superfine size and ultra-high coercivity.

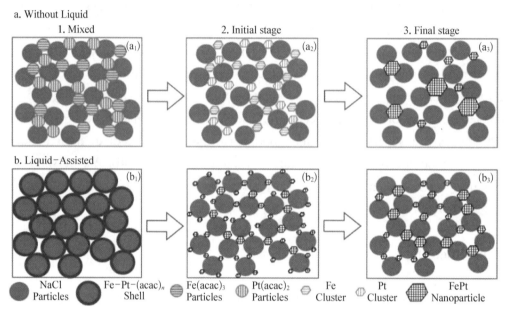

Fig. 1 The schematic diagram illustrates the formation process of $L1_0$-FePt NPs in the presence or absence of liquid-assisted

Metal-Organic Nanoprobe for Ratiometric Sensing Peroxynitrite and Hypochlorite Through FRET

Ding Zhaoyang[1,2], Wang Chunfei[1], Feng Gang[1], Zhang Xuanjun[1,*]

1 Faculty of Health Sciences, University of Macau, Macau SAR, China
2 Department of Chemistry, University of Washington, Seattle, USA
* Corresponding author: xuanjunzhang@um.edu.mo

【Introduction】

Various kinds of surface functionalization of the NMOFs could improve the stabilities and cell internalization, as well as grafting biomolecules, which is a key step towards biomedical applications. Protein, DNA, peptides, and polymers are usually applied to functionalize NMOFs. Dextran derived from the natural materials, with excellent biocompatibility and water solubility, have been used as matrix or functionalization material for NMOFs. As the surface functionalization material, dextran could control the interaction between nanoprobes and the analysts as biological media. Here, we reported a simple yet useful surface functionalization to construct ratiometric fluorescent nanoprobes for the detection of peroxynitrite ($ONOO^-$) and hypochlorite (ClO^-).

【Experimental】

In this work, the nanoprobe consisted of NMOF, surface functionalization material (dextran), and a small molecular probe (BPP). The aryl boronate group from BPP could react with $ONOO^-$ to form hydroxy derivatives and the phenothiazine group of BPP could be oxidizedinto phenothiazine-5-oxide after reacting with ClO^- (Scheme 1). Dextran was selected as an ideal agent to functionalize the surface of NMOF and combine it with BPP.

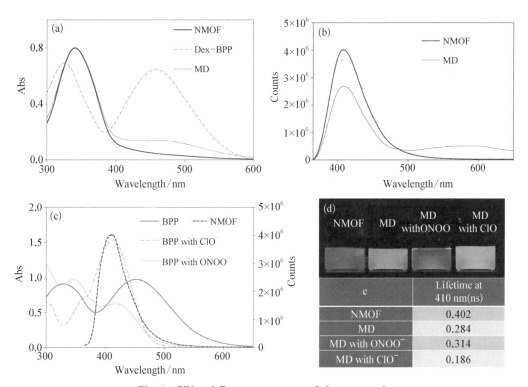

Fig. 1 UV and fluorescence curve of the nano-probe

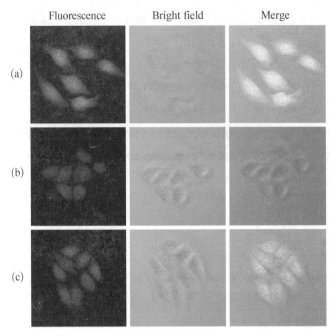

Fig. 2　Fluorescence images of Hela cells for $ONOO^-$ and ClO^-

【Results】

we have developed a novel metal-organic nanoprobe for ratiometric sensing $ONOO^-$ and ClO^- through FRET. Dextran is introduced to conveniently bind to the surface of the energy donor NMOF and combine with the energy acceptor molecular probe. This nanoprobe exhibits excellent photostability, high brightness, good water solubility, and cell internalization, which is successfully applied to the detection of $ONOO^-$ and ClO^- in living cells. These results suggest that this probe will be a promising tool in the investigations of $ONOO^-$ and ClO^- related diseases. The design strategy presented herein might be extended to construct other ratiometric fluorescent nanoprobes.

第三部分 冶金新技术
Section 3　Metallurgical Engineering

转炉炼钢底喷粉技术进展

朱荣[1,*], 胡绍岩[2], 李伟峰[1]

1 北京科技大学 2 苏州大学

* 通讯作者：zhurong1201@126.com

【引言】

随着转炉绿色洁净化生产的技术需求日益迫切,转炉底喷粉已成为近年来炼钢领域的热点技术,并被视为中国钢铁工业的绿色引领技术之一。相比于顶吹氧气、顶加块状石灰的常规转炉炼钢工艺,转炉底吹氧气-石灰粉具有吹炼过程平稳、原辅料消耗低、钢水纯净度高、固废产生量小等优点。但长期以来底喷粉转炉炉底寿命短的问题,从根本上制约了该技术的推广应用;本团队基于底喷粉转炉的炉底侵蚀机理,创新性地提出一种利用 CO_2 延长底喷粉转炉炉底寿命的新方法,并自主开发了成套的转炉底喷粉装备和工艺模型,在酒钢 120 吨转炉上成功开展了工业试验,本文将对工业试验的结果进行简要介绍。

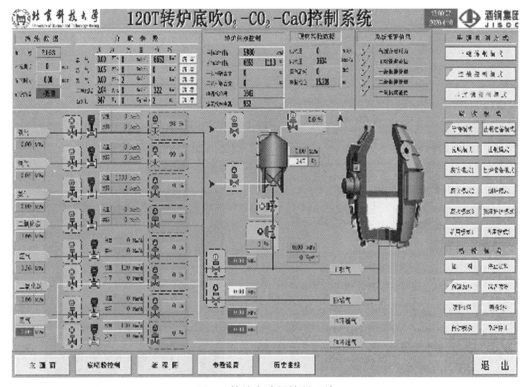

图 1 转炉底喷粉控制系统

【试验装置】

在转炉平台安装供气系统和喷粉系统,输粉管道穿过转炉耳轴,并沿炉壳到达转炉底部,通过粉气流分配器与安装在炉底的底吹元件进行连接,粉气流由底吹元件高速喷入转炉炉内,每支底吹元件的喷粉流量可达 100 kg/min;同时对顶吹氧枪的结构进行调整,系统优化了转炉的加料、吹炼、出钢和维护四项制度。

【结果】

经统计,底喷粉试验炉次的平均喷粉量为 1 000 kg/炉,石灰减总减量 800 kg/炉,折合吨钢石灰消耗量降低 7 kg,在此前提下,底喷粉炉次仍能获得良好的冶金效果。底喷粉炉次的平均磷含量为 0.010%,且终点磷含量波动范围更窄;在最新一阶段的试验炉次中,通过优化 CO_2 混入比例,平均磷含量仅为 0.007%,最低磷含量可达 0.003%。与此同时,钢水残锰含量由 0.13% 提高至 0.18%,最高锰含量可达 0.30%;钢水氧含量平均降低 112 ppm,出钢脱氧剂(钢砂铝)的消耗量由 295 kg/炉降低至 254 kg/炉。经初步核算,底喷粉炉次的吨钢成本降低约 27 元。

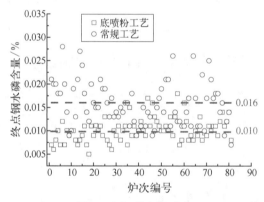

图 2 出钢磷含量对比

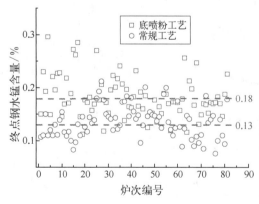

图 3 出钢锰含量对比

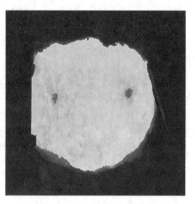

图 4 底吹喷嘴实拍图

Current Situations and Future Perspective of Ironmaking Industry and Green Ironmaking Technology

Zhang Jianliang*

University of Science and Technology, Beijing
* Corresponding author: jl.zhang@ustb.edu.cn

【Current situation of Chinese blast furnace ironmaking in recent years】

China is one of the first countries in the world to develop ironmaking technology. However, in the past country, China's ironmaking industry has been lagging for a long time compared with developed countries. The true development of China's ironmaking technology has been since the reform and opening up in 1979. The evolution of pig iron production in China and the World is shown in Fig.1. Chinese ironmaking process could be divided into three stages. The first stage is from 1978 to 2000, which is the stage of stable development. China's pig iron in this stage increased from 26 million tons to more than 100 million tons, successfully becoming the world's largest iron

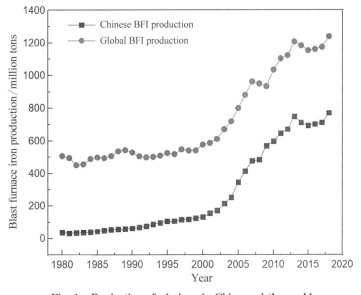

Fig. 1 Production of pig iron in China and the world

producer. The second stage, from 2000 to 2013, was a rapid development stage. During this time, China's pig iron production has maintained a rapid growth rate of 100 million tons per year (except for the financial crises of 2005 and 2009). The total pig iron production exceeded 700 million tons, accounting for more than half of the world's total production. The third stage is from 2013 to the present, where the ironmaking process in China has begun to shift to high quality. At present, due to the multiple pressures of resources and environmental protection, China's ironmaking industry is developing in an innovative direction. But in 2019, China's total iron production exceeded 800 million tons, accounting for more than 60% of the world's total iron production.

[Progress of Green ironmaking technology in recent years]

At present, global warming caused by excessively high levels of CO_2 in the Earth's atmosphere has attracted worldwide attention. The current level of CO_2 in the atmosphere has exceeded 400 ppm and is increasing year by year. The annual CO_2 emissions of the steel industry account for 6.7% of global total emissions, of which the emissions from ironmaking systems account for about 70%.

In 2018, the direct CO_2 emissions of China's steel industry reached 1.95 billion tons, accounting for about 34% of China's total CO_2 emissions, second only to the power industry. China's ironmaking industry is facing important challenges of energy-saving and emission reduction, and the CO_2 emission reduction of traditional ironmaking processes has almost reached its limit. Countries around the world are gradually launching new low-carbon ironmaking processes to reduce CO_2 emissions. The world's and Chinese steel industry has also accelerated the development of low-carbon ironmaking projects and hydrogen metallurgical projects in recent years.

In recent years, China has taken new steps in the new green ironmaking process. The main non-blast furnace ironmaking processes in China are shown in Fig. 2. The No.1 furnace of Baosteel Group's Bayi Iron and Steel Group was officially put into operation on June 18, 2015. Its prototype is the 2012 Corex-3000 ironmaking furnace of Baosteel, with a designed annual production of 1.5 million tons of molten iron. Also, Shandong Molong Petroleum Machinery Co., Ltd introduced the HIsmelt technology from Australia and successfully launched the furnace in June 2016. For the first time, HIsmelt's continuous industrial tapping was achieved. Direct smelting of iron using pulverized coal and fine ore has made a series of technological progress. The

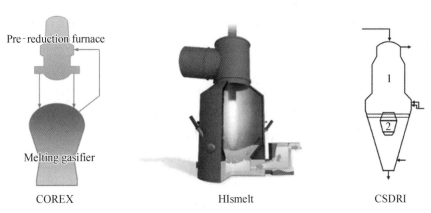

Fig. 2　The diagram of Chinese new ironmaking technology

HIsmelt process does not require coke ovens and sinter plants or pellet plants, so construction costs are reduced. HIsmelt technology can use the powder directly, without the need for a large mixing field, greatly reducing the footprint of the HIsmelt plant. A large amount of steam and surplus gas generated during the production process of the Shandong Molong HIsmelt smelt reduction process can be used for power generation, making its production system energy efficient. Moreover, in May 2013, China Shanxi Metallurgy Technology Co, LTD signed a contract with Iran MME to introduce advanced "direct reduction ironmaking" process technology and equipment (independently named CSDRI). With an annual output of 300 000 tons, this will be the first time that China has used coke oven gas to produce direct reduced iron. Gas-based direct reduction ironmaking is currently recognized as a more mature direct reduction ironmaking process in the world. Compared with traditional blast furnace ironmaking, the gas-based direct reduction has short process flow, low reaction temperature, low energy consumption, and pollutant emissions. The advantages of low product quality and high product quality will inevitably play an increasingly important role in future ironmaking production.

【Future perspective of Chinese ironmaking technology】

Despite the continuous development of various new green ironmaking processes, in the foreseeable future, the ironmaking process will still be dominated by the coking-sintering/pellet-blast furnace process. Pollutant emission from the coking-sintering-blast furnace ironmaking process accounts for approximately 90% of the steel process emissions. Energy consumption accounts for more than 60% of the total energy consumption of steel production, and production cost accounts for about 70% of the total cost of steel production. Therefore, if the ironmaking industry is to achieve sustainable development in the long term, energy-saving and emission reduction are the only way.

With increasing environmental and resource pressures, major steel companies will take hydrogen metallurgy as their long-term development goal. Compared with carbon metallurgy, the advantages of hydrogen metallurgy are (a) the reaction tail gas is a mixture of H_2 and H_2O, which can completely avoid CO_2 emissions; (b) Hydrogen can replace metallurgical coke, thereby reducing the environmental pollution. Compared with carbon metallurgy, the energy consumption of hydrogen metallurgy can be reduced by 57%, can CO_2 emissions can be reduced by 96%. Hydrogen metallurgy has become the primary choice for the world's steel industry to achieve CO_2 emission reduction from the source. Carbon metallurgy should still dominate in the next 20 years, but hydrogen metallurgy will completely replace carbon metallurgy in the future.

Technology of Fine Bubbles Generated by Argon Injecting into the Down-snorkel of RH Degasser

Liu Jianhua[1,*], Zhang Shuo[1], He Yang[1], Zhang Jie[2]

1 Institute of Engineering Technology, University of Science and Technology Beijing, Beijing 100083, China 2 School of Metallurgical and Ecological Engineering, University of Science and Technology Beijing, Beijing 100083, China
* Corresponding author: liujianhua@metall.ustb.edu.cn

【Introduction】

Inclusion removal can be effectively improved by dispersive argon microbubbles in the refining process of molten steel. Water modelling shows that the argon blown into the RH down snorkel could be sheared into fine bubbles by turbulent steel; then, dispersive microbubbles could be generated as the bubbles moved downward and spread in the molten steel.

【Experimental】

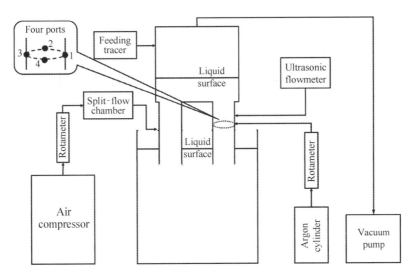

Fig. 1 The schematic of water model

【Results】

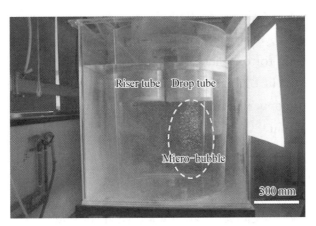

Fig. 2 The morphdogy of Mic robubble

Water modelling results show that fine bubbles can be generated in the water flow in down-snorkel and ladle during the RH refining by Ar injecting into the down-snorkel of RH degasser. Large circulate molten steel flow of RH refining, small Ar injection rate in down-snorkel and small orifice diameter are beneficial to the generation of fine bubbles. The mean diameters of bubbles generated by Ar injecting with orifices in diameters of 0.2 mm and 0.8 mm at the lifting gas flow rates of 3 m^3/h and 5 m^3/h are both 3.0 mm; while those with orifices in diameters of 1.5 mm are

3.1 mm and 4.5 mm respectively. Fine bubbles with diameters less than 5.0 mm are believed to be beneficial to inclusion removal.

High-speed Mold Width Adjustment Technology and Application of Double C

Xu Rongjun*

Central Research Institute of Baoshan Iron and Steel Co., Ltd

* Corresponding author: xurj@baosteel.com

【Introduction】

In this paper, the common types of mold width adjustment are compared, and the problems of parallel, zigzag, and step-by-step mold width adjustment are analyzed. This paper introduces the principle of width adjustment of double-C mold, which has the advantages of ensuring the close contact between the blank shell and the narrow edge copper plate as much as possible in the process of width adjustment; fast width adjustment speed (maximum 50 mm/min for one side of hot width adjustment, maximum 500 mm/min for one side of cold width adjustment), which is more suitable for the width adjustment of high casting speed; minimum stress on the blank shell; reducing the wear of copper plate, etc.; and gives the important parameters of double-C-type width adjustment.

【Experimental】

The cold width adjustment verification was carried out three times in total to verify the correctness of the program and the deviation between the theoretical setting value and the actual value. When there was an accident in the process of width adjustment, the width adjustment was terminated, and whether the mold could automatically recover the state before width adjustment. The verification process was smooth. Two thermal tests of mold width adjustment were carried out. The mold width adjustment speed was 1.2 m/min against y (up to 1 350 mm → 1 250 mm, width adjustment −100 mm); the mold width adjustment speed was 1.0 m/min with positive y (up to 1 250 mm → 1 350 mm, width adjustment ＋100 mm), and the width adjustment process was smooth. After the test, it was found that the actual process of adjusting the width of the slab was in good agreement with the theoretical calculation process. The surface quality of the slab was good, the side was flat without bulge or depression, the corner quality was good, and the actual length of the wedge-shaped slab with double C width adjustment was in accordance with the theory. The deviation between the slab of the Z-shape and double C width adjustment process model and the lower opening of the narrow edge copper plate in mold was analyzed and compared. The displacement, velocity, and acceleration images of Z-shaped and BIC width adjustment models are analyzed and compared. The relationship between the length of wedge transition blank.

【Results】

Double C width adjustment model: (1) The risk of steel leakage can be greatly reduced; (2) The width adjustment speed of the mold is from 15 m M/min at present, up to 50 mm/min; (3) The length of the transition wedge-shaped billet under the same conditions is reduced from generally more than 6 to 4 m, down by 50%; (4) The safety width adjustment each time is increased from 100 to 200 mm. (5) In the process of width adjustment, there is no need to reduce the speed (to adapt to the casting speed of 0.8 –1.8 m/min), which is beneficial to improve the surface quality of the slab (to prevent the level fluctuation and slag entrapment). This model has been applied in 5×5, 6×6 double-flow right angle mold and chamfered mold in Baosteel No. 2 steelmaking plant, and is being popularized to $4 \times$ casters in Baoshan base.

A Design of Self-generated Ti-Al-Si Gradient Coatings on Ti-6Al-4V Alloy Based on Silicon Concentration Gradient

Hu Xiaoyuan[1], Li Faguo[1,*], Shi Dongming[1],
Xie Yu[2], Li Zhi[1], Yin Fucheng[1]

1 Xiangtan University 2 China Baowu Steel Group Corporation Limited
* Corresponding author: lifaguo@xtu.edu.cn

【Introduction】

Titanium alloys are widely applied on aero-engines in the aviation industry, because of their high specific strength, low density, good corrosion resistance, good formability, excellent welding performances, high machinability, etc. However, titanium and titanium alloys have poor high-temperature oxidation resistance, which limits their applications on high-temperature components.

A method more effectively to improve the oxidation resistance of titanium alloys at high temperature is surface preparation coating, i.e. to prepare the coating with excellent oxidation resistance on the surface of titanium alloys. Aluminum alloy coating is used as an oxidation resistant coating. However, it has been proved that aluminum dipping on titanium alloy is easy to crack and flake off. Luckily, if silicon is added to the coating, the number of transverse cracks in the coating can be reduced. Thus, compared with the aluminum dipping layer, the Al-Si layer is less brittle and not easy to flake off.

The main intermetallic compounds in the Ti-Al system are Ti_3Al, $TiAl$, $TiAl_2$, and $TiAl_3$, among which only the $TiAl_3$ can form dense Al_2O_3 anti-oxidation protective film in the air to have good oxidation resistance. However, previous studies show that once the Ti-Al-Si coatings form on the titanium alloy's surface, the $TiAl_3$ phase layer is difficult to form. Therefore, we propose a method to form Si concentration gradient layers on the Ti-6Al-4V.

【Experimental】

The preparation method of Ti-Al-Si gradient coating is a combination of the hot-dipping method, casting infiltration method, in-situ self-generated method, and is thus referred to as the Self-generated Gradient Hot-dipping Infiltration (SGHDI) method. The mechanism of this combination can be explained in Fig. 1. During the hot-dipping, the alloy is immersed into the Al-Si melt in which an Al-Si coating is formed on the alloy surface through thermal diffusion. During the casting infiltration + in-situ self-generated, Al melt is poured into a cavity with SiO_2 on the wall of the mold to generate Si atoms when Al reacts with SiO_2 and then Al-Si concentration gradient alloys are formed when the self-generated Si atoms diffuse into the Al melt.

Self-generated Gradient Hot-dipping Infiltration (SGHDI) was proposed to prepare Ti-Al-Si gradient coatings on Ti-6Al-4V alloy. The coating samples with different dipping time (20, 30, 40, 50, and 60 min) were prepared at the dipping temperature of 800 ℃. The microstructure, phase structure, and composition of the coating observed by X-ray diffraction (XRD), scanning electron microscope (SEM),

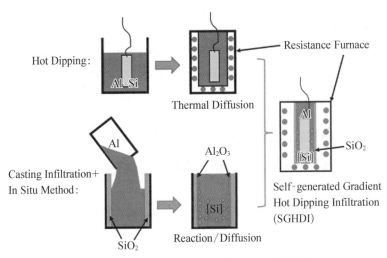

Fig. 1　Schematic diagram of process principle

and energy dispersive spectrometer (EDS).

【Results】

A new method of preparing coating referred to as the Self-generated Gradient Hot-dipping Infiltration (SGHDI) method was proposed in this paper. The conclusions were drawn as follows:

(1) The silicon concentration gradient coatings are in the sequence of Ti(Al, Si)$_3$ phase, τ_2 phase + L-(Al, Si) phase and L-(Al, Si) phase from the substrate.

(2) The Ti-Al-Si gradient coatings can provide good oxidation resistance to the substrate at a high temperature up to 800 ℃.

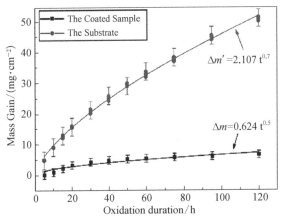

Fig. 2　Microstructure of Ti－Al－Si gradient coatings

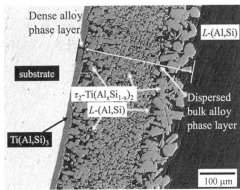

Fig. 3　The oxidation kinetic curves

(3) During high-temperature oxidation, three new alloy phase layers are formed between the substrate and the Ti(Al, Si)$_3$ layer. These phase layers (Ti$_3$Al, TiAl, and TiAl$_3$ + Ti$_5$Si$_3$) can prevent the cracks from propagating into the substrate, thereby blocking the diffusion of oxygen atoms into the substrate through cracks.

Nucleation and Growth Behavior of FePt Nanomaterials under High Magnetic Field

Wu Chun[1,2,*], Niu Zhiyuan[1], Wang Xiaoyang[2], Zhao Dong[2], Pei Wenli[2], Wang Kai[2], Wang Qiang[2,*]

1 Liaoning Technical University 2 Northeastern University;
* Corresponding authors: chun_wu@126.com; wangkai@epm.neu.edu.cn.

【Introduction】

Operating nucleation and growth process are the keys to control the crystal preparation, which is beneficial to obtain materials with controllable structure and performance. Normally, the nucleation and growth process is tuned by selecting the suitable composition or formation process parameters. It is no doubt that seeking some new strategies for operating the crystal growth is an effective way to break through existing technical obstacles, and obtains materials with novel structures or advanced functionality. Recently, the development of superconductive technology inspires us to apply the high magnetic field (HMF) into the materials formation process, to operate the nucleation and growth process of crystals by a non-contact and environmentally friendly style. The HMF presents significant effects on the nucleation and growth of 3D-bulk or 2D-film materials. Will, it affects the growth of low-dimensional materials?

【Experimental】

The wet-chemical synthesized FePt nanoparticles (NPs) were employed as the model materials for studying the effects of the HMF. The FePt NPs were synthesized by thermal decomposition reduction of Fe(CO)$_5$ and Pt(acac)$_2$. The 6 T HMF could be separately introduced into the nucleation stage, growth stage and both stages, because of the special burstnucleation of FePt NPs. The temperature and HMF strength curves that used for the synthesis of FePt NPs showed in Fig. 1. As the excitation rate was limited by 0.1 T/min, the normal curve was modified to generate or remove of the 6-T HMF. The most important characteristic of NPs, shape, and size, were collected from TEM images.

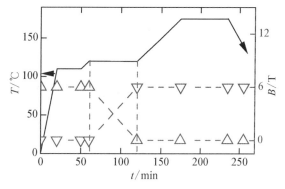

Fig. 1 Temperature and HMF strength curves

【Results】

Through carefully comparing the sizes of FePt NPs obtained by introducing the HMF in different stages, the effects of the HMF on nucleation and growth of FePt NPs were summarized in Fig.2. Application of the 6 T HMF at the nucleation stage, the grain sizes decreased about 10%, the nucleation rates increased about 50% and didn't show any effects on shapes. The HMF induced Lorentz force enhanced the diffusion and convection in the wet-chemical system, so the reduction reaction rates, location supersaturation, and nucleation rates of precursors were increased. While the 6 T HMF was applied at the growth stage, the grain sizes decreased about 20%, and the shape of FePt NPs transformed from cube to truncated-cube. The results highlight

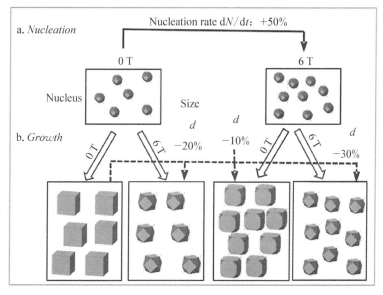

Fig. 2 Nucleation rates, shape and size of FePt NPs under HMF

the feasibility of controlling nucleation and growth of low-dimensional nanomaterials by application of the HMF.

Crystal Structure and Thermal Stability of the Intermetallic Phases in High-pressure Die Casting Mg-Al-RE Based Alloys

Yang Qiang*, Qiu Xin, Meng Jian

State Key Laboratory of Rare Earth Resource Utilization, Changchun Institute of Applied Chemistry, Chinese Academy of Sciences

* Corresponding author: qiangyang@ciac.ac.cn

【Abstract】

Although Mg-Al-RE (RE: rare earth) alloys have significant weight saving potential in automotive industries, their applications were interrupted due to their unsatisfactory mechanical performance. Intermetallic phases in structural magnesium (Mg) alloys are of practical significance for being able to optimize their microstructures for specific applications. Here we report a new alloy design concept that can develop an alloy for a given system with out-standing mechanical performance and low cost. Such a designed alloy exhibits even more excellent strength-ductility balance and cost creep-performance than the commercial/experimental die casting Mg alloys and A380 aluminum alloy. The alloy design concept used in this work is based on strictly controlling intermetallic phase components according to modifying alloy's compositions. Furthermore, the thermal stability of intermetallic phases in the Mg-Al-RE based alloys is of significant importance for their security services. Our study revealed that the well-known $Al_{11}La_3$ structure in a die casting Mg-4Al-2La-2Ca-0.3Mn alloy was decomposed when heat-treated at 400 ℃. Transmission electron microscopy characterizations reveal the decomposition process as Ca firstly segregated at Mg-$Al_{11}La_3$ interphase, then Al and Ca simultaneously segregated at the interphase boundaries and the radial Al_2Ca phase formed, finally the $Al_{11}La_3$ phase gradually decomposed into $Al_2(Ca, La)$ structure. First-principles calculations indicate that Ca atoms substituting for La atoms would decrease the stability of $Al_{11}La_3$ and result in that $Al_{11}La_3$ become metastable even at 0 K. Therefore, the decomposition off $Al_{11}La_3$

is attributed to that the interphase boundary segregation of Ca and the diffusion of Ca into $Al_{11}La_3$ change its thermodynamic stability.

【Results】

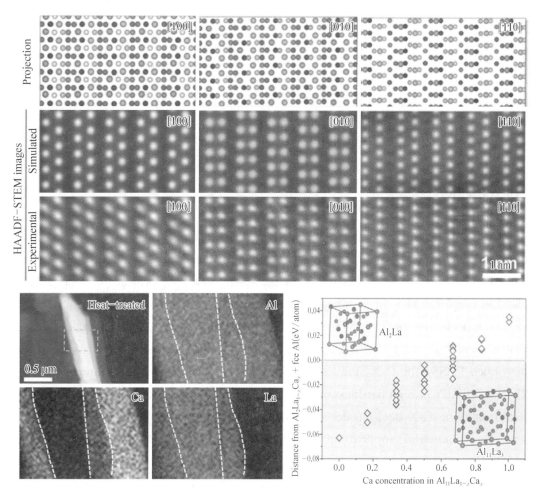

Fig. 1 New intermetallic phase in the Mg-Al-La alloy and the decomposition of the well-known $Al_{11}La_3$ phase in an Mg-Al-La-Ca-Mn alloy

Study on the Microstructure and Mechanical Performances of C-X Steel Processed by Selective Laser Melting (SLM) Technology

Chen Chaoyue[1]*, Yan Xincheng[1,2], Wang Jiang[1],
Ren Zhongming[1], Liu Min[1,2], Liao Hanlin[3]

1 Shanghai University 2 Guangdong Institute of New Materials 3 Univ. Bourgogne Franche-Comté

* Corresponding author: cchen1@shu.edu.cn

【Introduction】

Different from the traditional subtraction manufacturing, metallic additive manufacturing (AM) is a novel technology that joins materials to generate three-dimensional parts directly from CAD models based on the discrete-stacking principle. Owing to their excellent properties (e.g. high strength and toughness, good machinability, etc.), steels are very popular in the field of SLM manufacturing compared to other materials, such as titanium-based alloy, nickel-based alloy, aluminum-based alloy, etc. As a newly developed PHSS, the C-X stainless steel possesses extraordinary mechanical strength (UTS 1760 MPa), high hardness (51 HRC), and outstanding corrosion resistance after heat treatment. C-X steel has become a strong candidate for molding tools, tool parts, and other industrial applications requiring high strength and hardness. The C-X stainless steel was also invented as a substitute for SS grade 420. For the moment, the investigations on the microstructure features and mechanical properties of C-X stainless steel can be rarely found. Thus, in this work, the systematic studies were made on the microstructure characteristics and mechanical performance of SLM C-X stainless steel under different conditions. A profound discussion was made to enhance the understanding of the relationship between the underlying strengthening mechanism and microstructural evolution of SLM C-X steels.

【Experimental】

In our present work, the gas-atomized C-X SS powder (provided by EOS GmbH,

Germany) with spherical shapes was used as the feedstock materials. The cross-sectional observation with EDS mapping was given in the inserted image, which exhibits a uniform distribution of each element. The powder size distribution is $d_{10}=22.1~\mu m$, $d_{50}=35.4~\mu m$, and $d_{90}=56.9~\mu m$. To further adjust the microstructure and improve the mechanical properties, three different heat treatments (HT) were conducted on the AB samples: solution treatment at 900 ℃ for 1 h; aging treatment at 530 ℃ for 3 h; and solution treatment at 900 ℃ for 1 h followed by aging treatment at 530 ℃ for 3 h. All the specimens were air-cooled after withdrawn from the electric resistance furnace.

【Results】

High-performance C-X steel parts, a new type of PHSS, were successfully

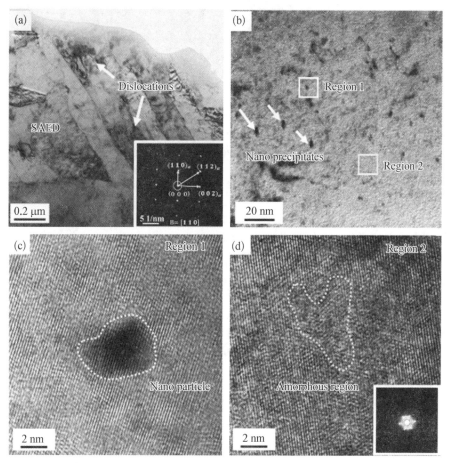

Fig. 1 BFI of the C-X steel under AB condition illustrating the existence of (a) parallel strip martensites with corresponding inserted SAED pattern, (b) nano precipitates (white arrowed), (c) nanoparticle of region 1, and (d) amorphous-nanocrystalline composite microstructure of region 2

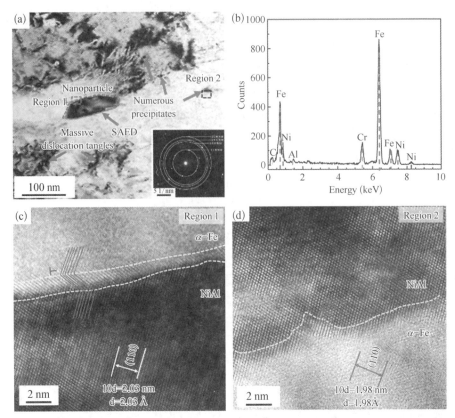

Fig. 2 Bright-field TEM and SAED images of aged specimens: (a) the numerous precipitates and massive dislocation tangles among the aged C-X steel matrix; (b) EDS results of the nanoparticle marked in (a); (c) high-resolution TEM (HRTEM) image of the partial coherence interface; (d) HRTEM of the coherence interface without elastic strain corresponding to the region 2 marked in (a)

manufactured by using SLM technology with optimized laser parameters. In this work, experiments were conducted to investigate the microstructure evolutions, precipitation hardening behavior, and mechanical properties (hardness and strength) affected by the different heat treatments. Based on the systematic investigative, the primary results are listed as follows: High-quality C-X steel parts with good weldability and formability can be successfully produced using SLM technology. A great number of fine martensite microstructures and a very small number of the reverted austenite constituted the matrix of the C-X steels, which displayed the complete martensite after solution treatment. This phenomenon can be observed using XRD and TEM technologies. Massive rod-like NiAl nano precipitates that were 3-25 nm in size and the amorphous phases generated by the tremendous cooling rate during the solidification of the laser molten pool were discovered in the AB sample. A similar phenomenon was detected in the TEM images of the aged sample, except that the scale of the NiAl intermetallic compounds (7 - 30 nm) was larger than that of the AB samples due to the growth of the nano precipitates during the aging treatment.

The solution-aging treatment increased the strength properties (hardness and ultimate tensile strength) from 350 HV0.2 and 1 043 MPa, respectively, to 510 HV0.2 and 1 610 MPa, respectively. This strengthening mechanism was discussed by using the Orowan mechanism and the calculated yield strength of the solution-aged samples (1 676 MPa) was similar to the measured results (1 528 MPa). The present work demonstrates that high-quality C-X steel with outstanding mechanical properties can be fabricated by SLM and subsequent heat treatment processes. These systematic experiments and theoretical analyses are useful for further theoretical research, industrial guidance, and large-scale production applications of the SLM generated C-X steels.

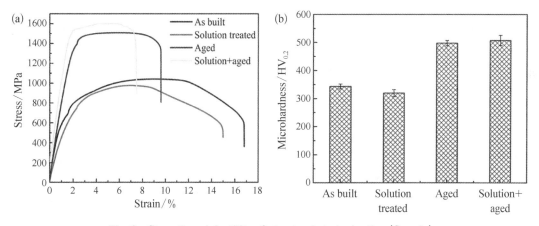

Fig. 3　Strength and ductility of structural steel: A - Rm (Sample)

金属凝固过程与均质化技术

翟启杰*,龚永勇,李仁兴,仲红刚

上海大学先进凝固技术中心,上海 200444

*通讯作者: qjzhai@shu.edu.cn

凝固是自然界中普遍存在的现象,更是金属制备的必经过程,因此认识凝固过程、掌握凝固规律,进而细化凝固组织、提高铸锭和铸坯的均质化水平是极为重要的研究课题。

长期以来，上海大学先进凝固技术中心（Center for Advanced Solidification Technology，CAST）一直致力于金属凝固过程及均质化技术的研究，先后提出并开发了金属凝固过程热模拟技术及相关装备，并自主研制了脉冲磁致振荡（Pulse Magneto-Oscillation，PMO）凝固均质化技术。

由于高温、不透明，加上工业生产中连续化、大规模和条件多变的特征，生产条件下金属凝固过程的试验研究一直是国际难题。翟启杰等提出凝固特征单元概念，并以特征单元热相似性为基础提出了以点见面的凝固热模拟新方法。在此基础上，针对冶金生产的迫切需求，发明了原位熔铸、温度场与枝晶生长调控和原位液淬等技术，研制了连铸坯凝固过程热模拟技术与装备；针对金属凝固共性问题，发明了差热分析与润湿角联测技术和动态加载诱导凝固裂纹等技术，研制出异质形核、凝固裂纹和亚快速凝固热模拟技术与装备。采用金属凝固热模拟技术，成功地将冶金生产条件下数吨至数十吨金属的凝固过程"浓缩"到实验室用百克金属研究，不仅可以揭示成分、尺寸、工艺制度等因素对凝固组织和元素分布的影响，还能获得固液界面形貌、界面前沿溶质和夹杂物演变等其他实验手段无法得到的重要信息，并认识夹杂物促进异质形核的能力、凝固裂纹形成的条件等冶金和铸造工作者长期关注的问题。CAST 依靠自主研制的热模拟装备，为宝钢等企业提供了大量技术服务，缩短了企业技术开发和工艺优化周期，降低了成本。图 1 示意了连铸坯枝晶生长热模拟技术的基本原理及装备。

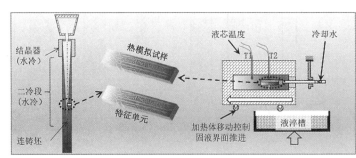 

(a) 连铸示意图　(b) 特征单元及热模拟试样　(c) 连铸热模拟技术原理　(d) 连铸坯枝晶生长热模拟试验机

图 1　连铸坯枝晶生长热模拟技术基本原理及装备

凝固组织细晶化和均质化是高性能金属材料发展的共性问题，对于钢铁材料来说，已经成为质量提升的技术瓶颈。许多有色金属可以通过化学法（变质处理、添加形核剂等）获得理想的凝固组织，但是钢铁材料一直未开发出便捷有效的细晶技术。经过十多年发展，CAST 揭示了脉冲电流细化金属凝固组织机制，并在此基础上提出原创性技术——PMO 凝固均质化技术。PMO 采用感应线圈处理金属凝固过程，在熔体内部感生出脉冲电流，为非接触式处理，对钢液无污染，处理时钢液平稳、操作便利、效果显著。其技术原理是感应脉冲电流通过"电致过冷"效应促进等轴晶大量形核并形成"结晶雨"，结晶雨堆积形成心部等轴晶区。该技术目前已经在方坯和矩形坯连铸中得到工业应用,经 PMO 处理的铸坯等轴晶面积明显增加,中心缩孔消除,均质化水平显著提高(见图 2)。

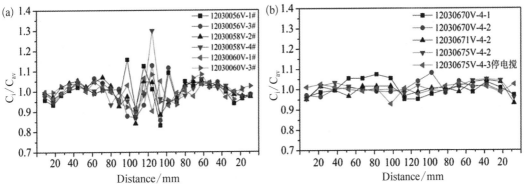

图 2 PMO 处理 GCr15 轴承钢连铸方坯的碳宏观偏析径向分布,(a) 未施加 PMO、(b) 施加 PMO

Modelling of the High Pressure Die Casting Process

Dou Kun[1,*], Ewan Lordan[1], Zhang Yijie[1], Alain Jacot[2], Fan Zhongyun[1]

1 Brunel Centre for Advanced Solidification Technology (BCAST), Brunel University London, Kingston Lane, Uxbridge, UB8 3PH, United Kingdom 2 Calcom ESI SA, Switzerland

* Corresponding author: Kun.Dou@brunel.ac.uk

【Introduction】

Numerical simulation is a powerful and cost-effective tool to optimize manufacturing processes whilst providing access to quantities that are difficult to obtain experimentally. Simulation of the high pressure dies casting (HPDC) process has been initiated in BCAST to assist in the design of experiments and investigate flow, heat transfer, microstructures, and defect formation in HPDC components during manufacturing. The project aims to understand the variability of mechanical properties in HPDC parts, which is a major concern to achieve weight reduction through lighter designs. To address this challenge in a cost-effective manner, a combined experimental and simulation-based approach has been developed.

【Experimental and Modelling】

A 40 kg crucible of A356 alloy after melt treatment is poured into the shot sleeve of a 4 500 kN cold chamber HPDC machine using a transfer ladle. The average temperature for melt shot sleeve, and die cavity is maintained at 680 ℃, 180 ℃ and 150 ℃ respectively. The molten metal is then injected into the die cavity at three sets of designed slow shot speed respectively combined with a constant fast shot filling speed of 3.6 m/s to produce eight-round tensile samples with a nominal gauge diameter of ϕ6.35 mm in accordance with ASTM standards. In the meantime, the entire HPDC process is modelled using the finite element (FEM) method to study the relationships between casting process parameters and casting defects formation.

【Results】

The thermal die cycling (TDC) process is modelled in modelled and validated against infrared camera measurements. (Fig. 1). On this basis, shot numbers before reaching steady-state could be determined and used for the actual HPDC process.

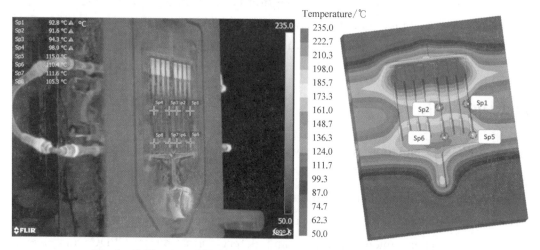

Fig. 1　Calibration of TDC process against infrared camera measurements

The influence of the piston profile on melt flow and defects formation in the shot sleeve and die cavity are studied systematically. The melt free surface evolution, air entrainment, and oxides distribution are considered. Three sets of piston profiles are selected with slow shot velocity changing from 0.2 – 0.3 m/s, 0.4 – 0.6 m/s and 0.6 – 1.0 m/s. Evolution of melt free surface in the shot sleeve, air entrainment in the shot sleeve, and defects formation in final cast tensile bars are modelled and compared respectively, as is shown in Fig. 2, proper slow shot piston velocity could be determined and suggested for real HPDC process, which helps in further understanding in variability for cast tensile bar mechanical properties.

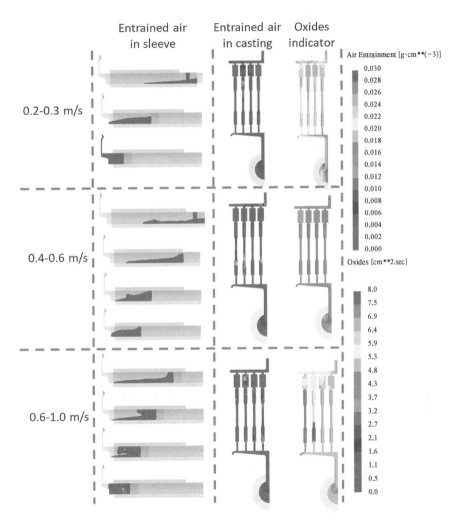

Fig. 2 Influence of piston profile on melt flow and defects formation

The formation and distribution of shrinkage porosity during the HPDC process is modelled considering the critical solid fraction for inter-dendritic melt flow and segregation of gaseous elements (main hydrogen in aluminium alloy) and its pressure drop in the mushy zone, a comparison between modelling results and Micro-CT scanning can be found in Fig.3.

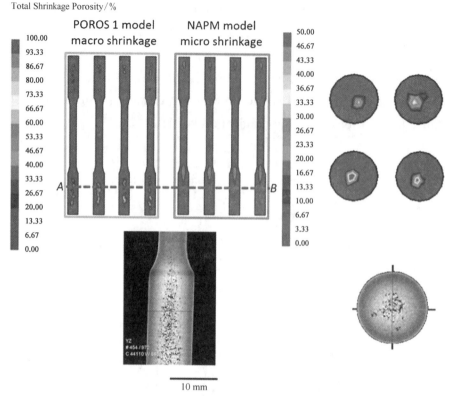

Fig. 3 Comparison of shrinkage porosity predictions (four tensile bars of different locations) against Micro-CT results

Some Considerations for the New Generation of High-efficiency Continuous Casting Technology Development

Zhu Miaoyong[1,2,*]

1 School of Metallurgy, Northeastern University, Shenyang, 110819, China
2 State Key Laboratory of Rolling and Automation, Northeastern University, Shenyang, 110819, China
* Corresponding author: myzhu@mail.neu.edu.cn

【Introduction】

Since continuous casting process with high productivity and high quality of steel play an important role to save energy and reduce consumption for the steel industry, high-efficiency continuous casting is the key to make the higher-efficiency steelmaking production line come true. The development connotation for the new generation of high-efficiency continuous casting technology should be higher productivity and higher quality, the core is higher casting speed and the objective is to have less cost, consumption, and emission. The solidification deflects such as crack, segregation, and porosity occurring frequently during the continuous casting, have restricted the development of continuous casting for high efficiency, and the new generation of continuous casting machine should have the advantages to prevent the occurrence of these solidification deflects. The new technologies developed and applied very recently for continuous casting mold, secondary cooling zone, and solidification final end represent the developing direction of continuous casting, and will be the standard technologies for continuous casting machines in the future.

【Technology】

Based on the formation mechanisms of corner crack and heat/mechanical behavior during the primary solidification of steel strand in mold, a new surface-crack-control technique to highly plasticize the solidification structure was developed. It includes an inner-convex-surface mold (ICS-Mold) to reduce the air gap between solidified shell and mold as much as possible and secondary cooling control technology at foot rollers for superfine grain (GSF-SSC). Then, for central segregation and porosity, a heavy reduction (HR) technique aiming at transferring the much more deformation to the mushy zone the in strand center at the solidification end was developed.

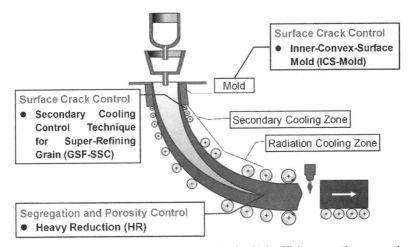

Fig. 1 Processing and equipment technologies for high-efficiency continuous casting

【Application】

The surface-crack-control technique has been applied on a large scale to 22 production lines in 13 large steel plants at home and abroad including Anshan Steel, Baosteel, Hegang, and Hyundai Steel. It covers a full range of strand shapes such as thin slabs, medium-thin slabs, conventional slabs, wide and thick slabs, and extra-thick slabs. After the application, the high-quality and efficient production of micro-alloy steel strands was guaranteed. The research group won the first prize of Liaoning Technical Invention in 2018 and the first prize of China Metallurgical Science and Technology in 2019. For heavy reduction, the research group developed a heavy reduction-enhanced compact sector (ECS) and a curving surface convex roll (CSC-Roll) and realized the reduction amount higher than 40 mm and 37 mm for the wide and thick slab and the bloom, respectively. Meanwhile, the research group developed the corresponding processes to ensure the reliable implementation of heavy reduction. The technique has been applied in steel plants at home and abroad including Baosteel, Pan Steel, Tang Steel, Ben Steel, and Hyundai Steel. The group has been awarded three first prizes issued by provinces and industry.

Fig. 2 Application of the present surface-crack-control technique

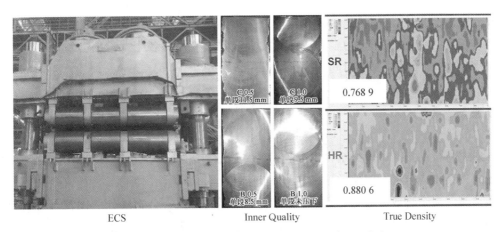

Fig. 3　Application of the present heavy reduction technique

连铸坯中夹杂物成分空间分布的预报

张立峰[1,*]，任英[2]，任强[2]，张月鑫[2]，王举金[2]，王亚栋[2]

1　燕山大学亚稳材料制备技术与科学国家重点实验室，河北　秦皇岛　066044
2　北京科技大学冶金与生态工程学院，北京　100083
* 通讯作者：zhanglifeng@ysu.edu.cn

【前言】

重轨钢对强度、韧性、洁净度和寿命的要求较高，而钢中非金属夹杂物，尤其是脆性夹杂物，会使其疲劳寿命明显降低。因此，控制重轨钢中非金属夹杂物的成分、尺寸和数量对于提高其使用寿命具有重要意义。目前已经有一些研究预测了连铸坯中夹杂物的尺寸和数量的分布，但是关于连铸坯中夹杂物成分分布的预测还鲜有报道。

【试验方法】

对一个典型重轨钢连铸坯，从内弧向外弧连续取样，应用自动扫描电镜分析钢中夹杂物的成分、尺寸、数量、形貌等信息。通过耦合凝固过程的传热、不同温度下夹杂物与钢平衡热力学以及钢中化学元素传质动力学，建立了一个连铸坯中夹杂物成分预报的模型（图1）。应用热力学方法计算出凝固冷却过程中夹杂

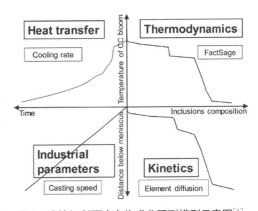

图1　连铸坯断面夹杂物成分预测模型示意图[1]

物成分随温度变化；应用传热和凝固模拟计算出连铸坯截面上任意位置的温度随时间和连铸坯长度的变化；应用动力学方法计算出夹杂物反应层中各个元素的含量随时间的变化，进而求出夹杂物成分随时间的变化。

【结果】

通过将自动扫描电镜分析得到的连铸坯从内弧到外弧钢中夹杂物的平均成分与模型预测的夹杂物成分进行对比，结果如图 2 所示。如图 2(a)所示，实际检测的夹杂物成分与模型预测的夹杂物成分较为一致，从连铸坯中内弧到外弧夹杂物成分变化较大的成分为 CaO 和 CaS。定义转化率为实际检测的 CaS 的含量与热力学计算得到的平衡 CaS 含量的比值，检测结果与预测结果对比如图 2(b)所示，检测结果与模型计算结果相一致。结果表明，应用此模型可以有效预测连铸坯中夹杂物的成分。

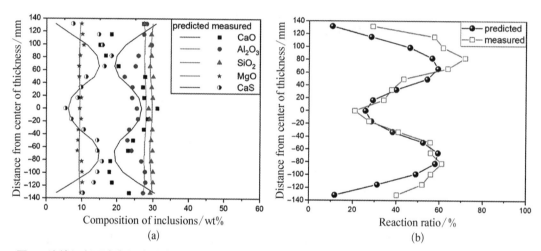

图 2　连铸坯断面夹杂物成分检测结果与模型预测结果对比：(a) 夹杂物平均成分；(b) 夹杂物转化率[1]

参考文献：

[1] Q. Ren, Y. Zhang, L. Zhang, J. Wang, Y. Chu, Y. Wang, Y. Ren. Prediction on the spatial distribution of the composition of inclusions in a heavy rail steel continuous casting bloom[J]. Journal of Materials Research and Technology, 2020, 9(3): 5648–5665.

机器学习模型预测钢铁冶金中的组织演变

牟望重*

瑞典皇家工学院材料科学与工程系

* 通讯作者：wmu@kth.se

【引言】

机器学习，广义来说，就是指让计算机具有像人一样的学习能力的一种特定模拟技术，从堆积如山的数据中寻找有用知识的数据挖掘技术，也通常地称为"大数据"。通过运用机器学习技术，可以根据用户的购买记录向客户推荐相关产品，抑或从视频库中寻找其喜欢的视频资料等。

根据机器学习所展示的特点，其冶金工业以及材料设计的各个研究领域中都有着较强的适用性，近年来成为研究的新热点，通常也称为"冶金大数据"技术。与其他材料设计模拟手段类似，机器学习也是基于大量可靠数据构成数据库，来通过冶金过程工艺参数预测最终组织、性能。多年前，人工神经网络等模拟技术已经开始在冶金中进行应用。近年来，随着计算机以及模拟技术的发展以及在数字化智能制造的大背景下，机器学习技术在冶金中有着更为广泛的发展，本文以钢铁材料的组织演变为例简要介绍机器学习在冶金中的应用。

【方法论】

机器学习，根据所处理的数据类型的不同，可以主要分为监督学习（Supervised learning）和无监督学习（Unsupervised learning），在特定情况下也包括强化学习。具体机器学习分类方法如图1所示。监督学习，是指有计算机有目的性地从周围的环境中获取知识、信息，根据训练过程中的所获得的经验对没有学习过的问题给出正确的解答，使计算机获得这种广泛化的能力。对于这一类型主要包括预测数值型数据的回归（Regression）、预测分类标签的分类（Classification）等。无监督学习，是指在没有从周围环境中获得可靠性参考的情况下，计算机的自我学习过程，通常的应用为计算机在互联网中自动收集信息，并从中获取有用信息。无监督学习可以不仅仅局限在解决有明确答案的问题，因此，它的学习目标可以不是十分明确，一个典型的应用为聚类（Clustering）。

线性模型（Linear model）作为研究机器学习模型的基础，它是从学习对象 f 函数是一维的情况下开始说明的。在对应函数 f 进行近似时，最简单的模型就是线性模型 $\theta \times$

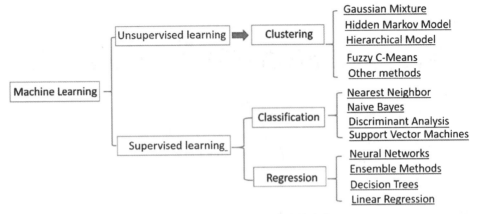

图 1　不同机器学习建模技术的分类

x。通过这个参数学习,完成函数的近似计算。由于这个模型只能表现出直线关系的输入输出函数,所以在解决实际问题方面,往往没有很大的实用价值。从个人观点来看,有监督过程的分类和回归是目前钢铁冶金中目前应用较为广泛的方法。本文中先进钢铁材料的组织预测,应用较为广泛的为有监督回归中的集成学习(Ensemble learning)。它是指把性能较低的多种弱学习期,通过适当组合而形成的强学习期的方法。近年来,集成学习的研究一直是机器学习领域研究的热点之一。目前,集成学习方法主要有两种,一种是对弱学习器独立进行学习的 Bagging 学习法,另一种是对多个弱学习器依次进行学习的 Boosting 学习法。集成学习不仅适用于回归分析,也适用于分类等各种类型的机器学习任务,本文中的实例仅限于回归分析。集成学习回归分析的训练过程如图 2 所示,主体过程可分为数据库准备(收集、清理等)、模型训练、模型预测等主要过程。

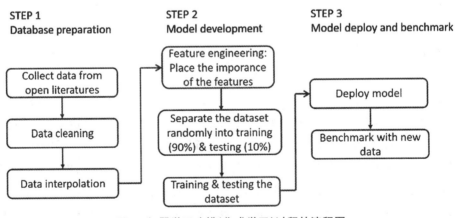

图 2　机器学习建模(集成学习)过程的流程图

【结果】

图 3 展示了应用机器学习预测马氏体相转变开始温度(Ms)的例子,该数据库含有

2 500 个不同钢种所对应的 Ms 相变温度,基于模型精确度考虑,只有低碳低合金钢在考虑范围之内,保证奥氏体化温度时不含其他杂质相,忽略其他因素对 Ms 相变温度的干扰。从图 3(a)可见,Ms 随 C 数量增多而降低,此预测结果完全符合物理冶金常识。图 3(b)显示了预测值与实际值的偏差。该模型的预测能力与预测 Ms 的热力学模型相比较,可得到略好于现存热力学模型的精度值。

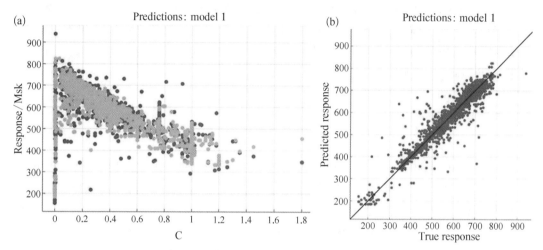

图 3　应用机器学习预测马氏体相转变温度(Ms)的例子

高温长时蠕变后 G115 钢微观组织的演变及其对硬度的影响[①]

周任远[1],朱丽慧[1,*],柯志刚[1],王金磊[1],翟国丽[2]

1　上海大学　2　宝钢中央研究院

* 通讯作者:lhzhu@i.shu.edu.cn

【引言】

G115(9Cr-3W-3Co-1CuVNbB)是钢研院研发的新一代马氏体耐热钢,是 650 ℃ 超超临界火电机组用马氏体耐热钢的最佳候选材料。G115 以 W 代 Mo,添加 Co 和 Cu 元素,其高温持久强度较 T92 有很大程度的提高。耐热钢高温蠕变后微观组织演变对高温性能有至关重要的影响,有必要研究 G115 钢高温长时蠕变后微观组织的演变以

---

① 基金项目:国家重点研发计划(2016YFC0801901)

及对性能的影响。

【实验方法】

对 G115 钢进行 650℃ 中断蠕变试验,选择蠕变 1 000 h、1 879 h、5 000 h、7 288 h 以及 G115 的正回火态试样进行研究。利用扫描电镜(SEM)和透射电镜(TEM)重点观察析出相、位错和马氏体板条在高温蠕变过程中的演变,并统计析出相的平均尺寸和体积分数、位错密度和马氏体板条宽度。最后通过强度计算来讨论高温蠕变过程中微观组织对 G115 强度、硬度的影响。

【结果】

通过 SEM 和 TEM 观察可知,G115 钢在 650℃ 蠕变后析出 Laves 相、$M_{23}C_6$ 相、富铜相和 MX 相。Laves 相和 $M_{23}C_6$ 相尺寸较大。大部分 Laves 相和 $M_{23}C_6$ 相在原奥氏体晶界和马氏体板条界上析出,也有少量在晶内析出。富铜相和 MX 相较为细小,大多在晶内析出。析出相、位错和马氏体板条随蠕变时间的具体变化如图 1 所示。随着蠕变时间的延长,Laves 相、$M_{23}C_6$ 相尺寸增大,而 MX 相尺寸基本保持不变。富铜相也有增大趋

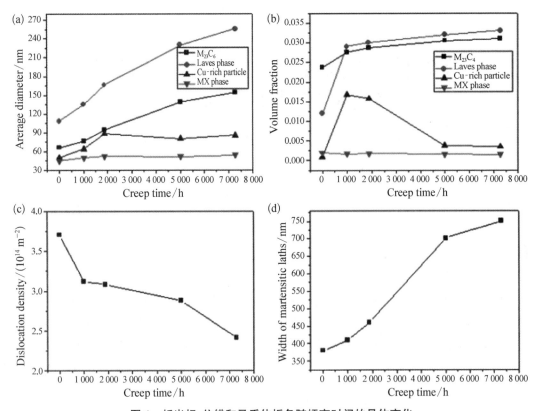

图 1 析出相、位错和马氏体板条随蠕变时间的具体变化

势,并且部分富铜相在蠕变过程中发生溶解。蠕变初期,位错密度下降较快但马氏体板条宽化较缓。蠕变 1 000~5 000 h,位错密度下降减缓但马氏体板条宽化加快。蠕变 5 000 h 后,位错密度进一步下降而马氏体板条宽化速率变缓。

$$\sigma_{\text{Or}} = \frac{0.3Gbf^{1/2}}{d}\ln\left(\frac{d}{2b}\right) \tag{1}$$

$$\sigma_\rho = M\alpha_1 Gb\sqrt{\rho} \tag{2}$$

$$\sigma_L = \frac{\alpha_2 Gb}{\lambda} \tag{3}$$

$$\sigma_{\text{ss}} = K_i C_i^{3/4} \tag{4}$$

析出强化、位错强化、马氏体板条强化和固溶强化可分别利用公式(1)、(2)、(3)和(4)计算(其中 G 是切变模量,b 为柏氏矢量,d 和 f 分别为析出相的平均尺寸和体积分数;M 是泰勒常数,ρ 为位错密度;λ 为马氏体板条宽度,α_1 和 α_2 为常数;K_i 为固溶强化系数,C_i 为固溶元素的原子百分比。本文主要考虑 W 和 Co 的固溶强化),结果如图 2(a)所示。

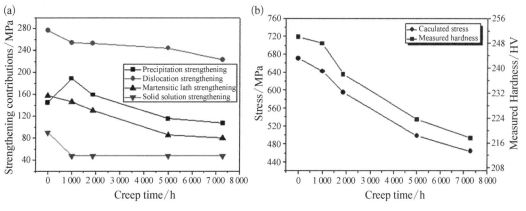

图 2　总强度的计算结果及其与测量的硬度的比较

对比总强度的计算结果和测量的硬度,发现两者的变化趋势吻合得较好。在蠕变初期,随着 Laves 相的大量析出,基体中 W 含量急剧下降,导致固溶强化快速降低。但析出强化的快速升高弥补了部分位错强化、板条强化和固溶强化的下降,导致强度和硬度下降较缓。蠕变 1 000~5 000 h,位错强化和固溶强化的变化较为平稳,但析出强化和板条强化下降较快,因而强度和硬度快速下降。蠕变 5 000 h 后,位错密度的持续下降使得强度和硬度进一步降低。尽管 G115 钢中析出相产生的强化效果不是最大的,但一方面,析出相可以减缓位错密度的降低和马氏体板条的宽化,增强位错和马氏体板条强化效果。另一方面,析出相产生的析出强化对 G115 钢蠕变过程中强度和硬度的变化有较大的影响。因此,析出相对提升 G115 钢的性能至关重要。

The Formation Mechanism of Dislocation Patterns under Low Cycle Fatigue of High-manganese Austenitic TRIP Steel with Dominating Planar Slip Mode

An Dayong*, Stefan Zaefferer

Max-Planck-Institut für Eisenforschung, Max-Planck-Str. 1, 40237 Düsseldorf, Germany

*Corresponding author: andyallone@icloud.com

【Introduction】

Austenitic high-manganese steels (HMnSs) with transformation-induced plasticity (TRIP) exhibit a good combination of strength and ductility under monotonic deformation, while their fatigue behavior has been rarely studied. Nevertheless, many of the HMnSs are used for automotive applications, where they undergo cyclic deformation during service life. Fatigue is, thus, an important damage mechanism for HMnSs, and understanding their fatigue behavior is of great importance to design reliable advanced materials.

【Experimental】

Digital image correlation (DIC) was performed during cyclic loading to measure the local strain amplitude of the fatigued specimen. ECCI at interrupted cycles was used to capture the early dislocation pattern evolution process. Cross-correlation based electron backscattering diffraction (CC-EBSD) with the high angular resolution has been applied in this study to measure the residual strain/stress distribution and geometrically necessary dislocations (GNDs) of the fatigued polycrystalline materials.

【Results】

Heterogeneous dislocation structures in planar slip material are observed, i.e. regularly spaced dislocation walls with thicknesses of $0.2 - 0.4\ \mu m$, as shown in

Fig.1(a), separated by channels of low dislocation density with widths of around 1 μm. Besides, the residual stress distribution is closely related to the dislocation arrangement, see Fig.1 (b). Furthermore, the dislocation walls are found to be mainly composed of edge type dislocations. In detail, edge type dislocations (1 1 1)[1 0 −1] and (−1 1 −1) [−1 0 1] are probed by CC-EBSD [Fig.1 (c)]. The colour scale in Fig. 1 (c) indicates positive (⊥, red) and negative (⊤, blue) signs of the Burgers vectors. We experimentally observe that the interactions between dislocations of same Burgers vectors gliding on intersecting slip systems, i.e. collinear interaction, lead to the annihilation of the screw parts and the formation of edge dislocation dipoles, which facilitate the construction of edge dislocation walls in material deforming in planar slip mode.

Theory and Application of High-value Utilization of Typical Bulk Industrial Solid Waste

Zhang Yuzhu[1,*], Xing Hongwei[1], Hu Changqing[1],
Zhao Kai[1], Chen Wei[1], Wang Baoxiang[2]

1 North China University of Science and Technology 2 Tangshan Shunhao Environmental Protection Technology Co., Ltd.

* Corresponding author: zyz@ncst.edu.cn

【Introduction】

The domestic power industry emits about 700 million tons of fly ash each year, and with the development of the power industry, it is expected to reach 900 million tons in 2024; the steel industry produces about 250 million tons of blast furnace slag every year, and carry sensible heat equivalent to about 15 million tons of standard coal. Both the above-mentioned typical bulk industrial wastes contain large amounts of inorganic oxides such as CaO, Al_2O_3, SiO_2, MgO. To solve the problems of high-value utilization and sensible heat utilization of blast furnace slag, fly ash, etc., based on the team's strong theoretical research foundation in silicate melt high-temperature mineral phase reconstruction and high-speed spinning, systematic development of key technologies for online continuous quenching and tempering of blast furnace slag, supplemental heating, and fiber formation; Research and development of key technologies for fly ash high-temperature molten ore phase reconstruction and ultra-fine ceramic fiber preparation.

【Experimental】

Based on the principle of "using waste to waste", through the establishment of multi-phase diagram theoretical analysis, a theoretical system of slag quenching and tempering is established, on this basis, two solid wastes with high SiO_2 iron tailings and fly ash are selected as blast furnace slag modifying agent. Through FactSage thermodynamic simulation, constant temperature cooling crystallization behavior, continuous cooling crystallization behavior experimental research, and XRD analysis, study slag cooling process crystallization behavior; Through numerical simulation and measurement and analysis of physical and chemical properties of high-temperature melt, the melting and slagging behavior of modifying agent particles and the change of system heat enthalpy were studied; Through the determination of the viscosity of quenched and tempered slag, the boundary conditions and time of slag homogenization are determined. Combined with the boundary layer theory and the centrifugal fiberization experiment, the blast furnace slag centrifugal fiberization process was determined.

【Results】

Research on continuous preparation of mineral wool fiber by online quenching and tempering blast furnace slag was carried out: Established a theoretical system for online quenching and tempering of blast furnace slag, which provides a theoretical basis for the direction of slag quenching and tempering and the choice of modifying agent; an unsteady three-dimensional diffusion model of heat transfer and diffusion of modifying agent molecules in high-temperature slag was established. A coordinated control chart of slag system temperature and slag fluidity temperature was developed. Developed a dynamic control method for the heating and homogenization of quenched and tempered slag; Combined with the boundary layer theory, the mathematical model and fluctuation model of the thickness distribution of the boundary layer of the slag on the roll surface were established, and the centrifugal fiber forming process of quenched and tempered slag was developed; The production line of fiber insulation board with an annual output of 20 000 tons was built in the plating plant of HBIS Group Tangsteel Company, and the development of fibrous photocatalytic materials loaded with TiO_2, plate-shaped, tubular insulation materials and sound insulation materials, and fiber-reinforced cement concrete a series of high value-added products, greatly expanded the application field of blast furnace slag fiber. Developed

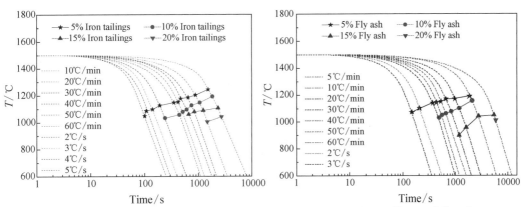

Fig. 1　Continuous crystallization behavior of slag after adding iron tailings and fly ash

key technologies of high-temperature dynamic melting and upgrading of fly ash and blast furnace slag based on variable-frequency resistance furnace, and ultra-high-speed spinning of high-temperature melt to prepare ultra-fine ceramic fibers. The automatic production line with a single-line production capacity of 25 000 tons has an annual output of 150 000 tons of various new ceramic fiber products.

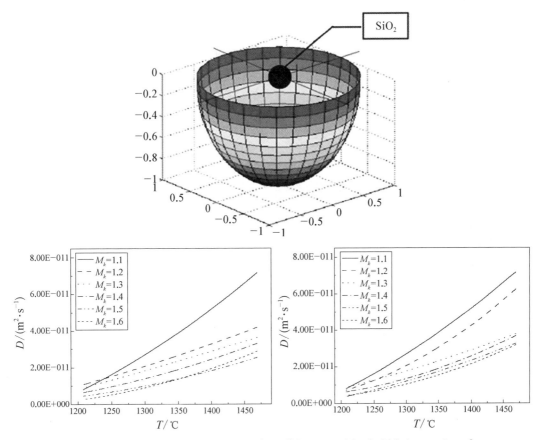

Fig. 2　Melting and diffusing behavior of conditioner particles in high-temperature slag

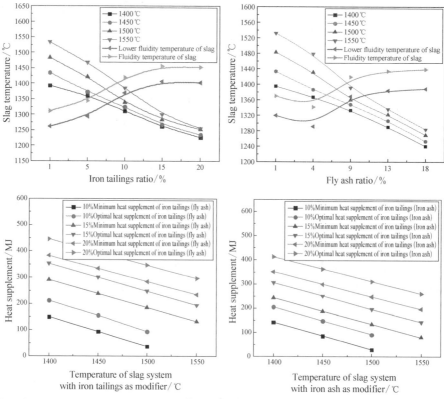

Fig. 3　Coordinated control diagram of slag system temperature and slag fluidity temperature

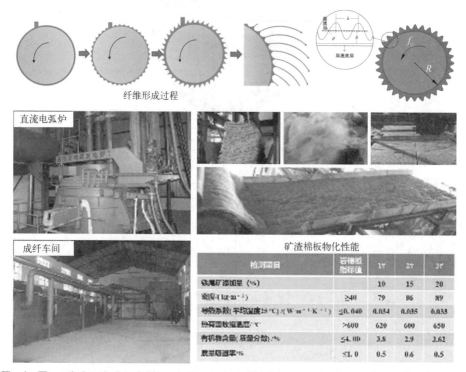

Fig. 4　Theoretical analysis and demonstration project of high temperature slag continuous fiber formation

第四部分 功能材料

Section 4 Functional Materials

Effect of Nanoparticles Al$_2$O$_3$ and Rare Earth on New Alumina Dispersion-strengthened Copper Alloy

Fu Yabo[1,*], Pan Qingfa[1], Li Shufeng[2], Huo Yanqiu[1]

1 Zhejiang Provincial Key Laboratory for Cutting Tools, School of Pharmaceutical and Materials Engineering, Taizhou University, Taizhou 318000, P.R. China   2 School of Materials Science and Engineering, Xi'an University of Technology, Xi'an 710048, P.R. China
* Corresponding author: Yabo Fu, Lgdfyb@163.com

【Introduction】

Considerable researches have been conducted on the preparation and properties of ADSC alloys for their characterization. Guo has studied the effect of cold working on the microstructures and tensile fracture behavior of ADSC alloy. Fu has studied a new type of ADSC alloy containing rare-earth metals Ce that was prepared via a sol-gel and powder sintering method. Zhang has studied that intermetallic compounds Cu-Al-RE (RE: rare-earth metals La and Ce) containing the same weight percentage of Al and RE showed the lowest enthalpy; therefore, these ternary compounds were considered to be more stable than the other rations compounds. However, the behavior and reaction mechanism of the strength and electrical conductivity on ADSC alloy with the addition of La and Ce have seldom been reported.

In the paper, we aim to study the strength and electrical conductivity behavior in situ reaction using rare-earth La and Ce. Solving the clustering on nanoparticles Al$_2$O$_3$ under spark plasma sintering (SPS) and the addition of La and Ce for ADSC alloy is the significant difference. A new ADSC alloy containing rare-earth nanoparticles will be fabricated, whose strength, hardness, electrical conductivity, and density are determined.

【Experimental】

The specimens prepared by spark plasma sintering (SPS) were labeled as below:

Cu-0.2 wt% Al_2O_3 (A), Cu-0.2 wt% Al_2O_3-0.2 wt% La (B), Cu-0.2 wt% Al_2O_3-0.2 wt% Ce (C), Cu-0.7 wt% Al_2O_3 (D), Cu-0.7 wt% Al_2O_3-0.7 wt% La (E), and Cu-0.7 wt% Al_2O_3-0.7 wt% Ce (F). Intermetallic compounds Cu-Al-RE (RE: rare-earth metals La and Ce) containing the same weight percentage of Al and RE due to the lowest enthalpy. Therefore, these ternary compounds were considered to be more stable than compounds with other ratios for the constituents, in accordance with a previously published report. Sizes of Al_2O_3 are 20 – 30 nm in width and 40 – 60 nm in length. The fabrication procedures were as flow chat: mixed powders of Cu ＋ Al_2O_3 ＋ RE → absolute alcohol→planetary ball milling at 180 rpm for 5 h → SPS at 860 ℃ for 10 min for ϕ30 mm→hot extrusion at 800 ℃ for ϕ9 mm.

【Results】

Fig. 1 shows the tensile strength and electrical conductivity after hot extrusion, from Fig.1 (a) we can see that the addition of 0.2 – 0.7 wt% La and 0.2 – 0.7 wt% Ce can increase the tensile strength, especially rare-earth Ce have a more strengthening affection. From Fig.1 (b) we can see that addition of 0.2 – 0.7 wt% rare-earth La can decrease the electrical conductivity, however, addition of 0.2 – 0.7 wt% Ce can crease the electrical conductivity. So, Cu-0.7 wt% Al_2O_3-0.7 wt% Ce has the highest tensile strength in all specimens, the electrical conductivity is higher than that of Cu-0.7 wt% Al_2O. Rare-earth Ce has a key factor in improving the properties of the new ADSC alloy.

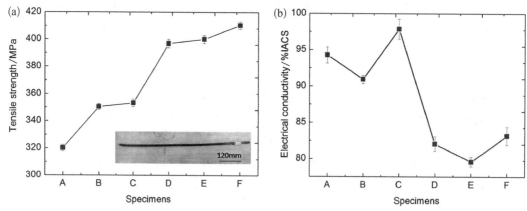

Fig. 1 Tensile strength and electrical conductivity curves: (a) tensile strength, (b) electrical conductivity. (A) Cu-0.2 wt% Al_2O_3, (B) Cu-0.2 wt% Al_2O_3-0.2 wt% La, (C) Cu-0.2 wt% Al_2O_3-0.2 wt% Ce, (D) Cu-0.7 wt% Al_2O_3, (E) Cu-0.7 wt% Al_2O_3-0.7 wt% La, and (F) Cu-0.7 wt% Al_2O_3-0.7 wt% Ce

Fig.2 shows the fracture morphologies of tensile for Cu-0.7 wt% Al_2O_3 and Cu-0.7 wt% Al_2O_3-0.7 wt% Ce, from which we can see that adding with the rare-earth Ce, there are no cracks in the fracture morphologies. However, without rare-earth Ce, there are many big cracks in fracture morphologies, resulting in low tensile strength.

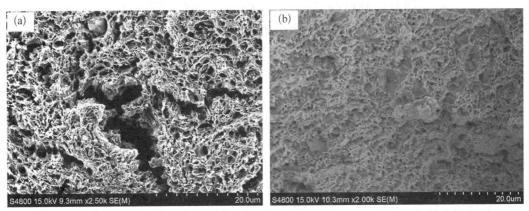

Fig. 2　Fracture morphologies of tensile: (a) Cu-0.7 wt% Al_2O_3, (b) Cu-0.7 wt% Al_2O_3-0.7 wt% Ce

【Conclusions】

1. Cu-0.7 wt% Al_2O_3 added with 0.7 wt% Ce which is fabricated with SPS shows the high tensile strength and the high electrical conductivity due to excellent dispersed element Ce.

2. Nanoparticles Al_2O_3 encircled with nanoparticles CeO_2 and Cu_2O, which are generated in situ are evenly distributed in the Cu matrix, resulting in increased tensile strength. Lattice distortion does not occur in the new ADSC alloy. Cu atoms are replaced by Al generated in situ to form a solid solution with a face-centered cubic structure owing to the similar atomic radii and electronegativities for Cu and Al, resulting in increasing the electrical conductivity.

Focusing Nano Metals Under Pressure

Dong Hongliang*, Xu Jianing, Zhou Xiaoling,
Chen Zhiqiang, Zhang Hengzhong, Chen Bin

Center for High Pressure Science and Technology
Advanced Research, Shanghai 201203, China
* Corresponding author: hongliang.dong@hpstar.ac.cn

【Introduction】

Pressure, as a fundamental thermodynamic variable, can drastically change the physical and chemical properties of nanoscale materials[1,2]. We could carry out high-pressure research and in situ characterization via a diamond anvil cell (DAC) or in-situ TEM. Various interesting properties were observed compared to chemical pressure[3]. Here, we would summarize and review the following exciting examples, especially metallic materials, achieved under high pressure[4-12]. Then we could see the realization of opening up the bandgap of a few-layer graphene[9] and the enhanced mechanical properties of nanometals under high pressure[5].

【Experimental】

As shown in Fig.1, high-pressure radial DAC and axial DAC were used to measure the structure evolution and properties.

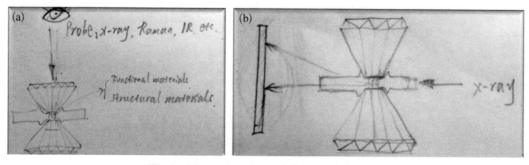

Fig. 1 Diamond anvil cell (a) axial, (b) radial

【Results】

Transport properties and mechanical properties of materials could be effectively tuned by pressure. The discovery of more interesting phenomenons of metallic materials, especially nano metals, will broaden the horizon in materials design and device applications to a large extent.

References:

[1] Ho-Kwang Mao, Xiao-Jia Chen, Yang Ding, Bing Li, and Lin Wang. Solids, liquids, and gases under high pressure[J]. Rev. Mod. Phys., 2018, 90: 015007.
[2] Yu Deng, Ruopeng Zhang, Thomas C. Pekin, Christoph Gammer, Jim Ciston, Peter Ercius, Colin Ophus, Karen Bustillo, Chengyu Song, Shiteng Zhao, Hua Guo, Yunlei Zhao, Hongliang Dong, Zhiqiang Chen, Andrew M. Minor. Functional Materials Under Stress: In Situ TEM Observations of Structural Evolution[J]. Adv. Mater. 2019: 1906105.
[3] Youwei Xiao, Yixuan Wu, Pengfei Nan, Hongliang Dong, Zhiwei Chen, Zhiqiang Chen, Hongkai Gu, Binghui Ge, Wen Li and Yanzhong Pei. Cu interstitials enable carriers and dislocations for thermoelectric enhancements in n − $PbTe_{0.75}Se_{0.25}$[J]. Chem, 2020, 6: 523 − 537.
[4] Yunlei Zhao et al. Journal of Nanoscience and Nanotechnology, 2020, accepted.
[5] Xiaoling Zhou, Zongqiang Feng, Linli Zhu, Jianing Xu, Lowell Miyagi, Hongliang Dong, Hongwei Sheng, Yanju Wang, QuanLi, Yanming Ma, Hengzhong Zhang, Jinyuan Yan, Nobumichi Tamura, Martin Kunz, Katie Lutker, Tianlin Huang, Darcy A Hughes, Xiaowu Huang, Bin Chen, High-pressure strengthening in ultrafine-grained metals[J]. Nature, 2020, 579: 67 − 72.
[6] Yu Deng, Christoph Gammer, Jim Ciston, Peter Ercius, Colin OPhus, Karen Bustillo, Chengyu Song, Ruopeng Zhang, Di Wu, Zhiqiang Chen, Hongliang Dong, Armen G Khachaturyan and Andrew M Minor. Hierarchically-structured large superelastic deformation in ferroelastic-ferroelectrics[J]. Acta Materialia 2019, 181: 501 − 509.
[7] Resta A Susilo, Guowei Li, Jiajia Feng, Wen Deng, Mingzhi Yuan, Shujia Li, Hongliang Dong and Bin Chen. Pressure-induced structural and electronic transitions of thiospinel Fe_3S_4[J]. J. Phys. Condens. Matter. 2019, 31: 095401.
[8] Yanwei Huang, Yu He, Howard Sheng, Xia Lu, Haini Dong, Sudeshna Samanta, Hongliang Dong, Xifeng Li, Duck Young Kim, Ho-kwang Mao, Yuzi Liu, Heping Li, Hong Li and Lin Wang. Li-ion battery material under high pressure: amorphization and enhanced conductivity of $Li_4Ti_5O_{12}$[J]. National Science Review, 2019, 6: 239 − 246.
[9] Feng Ke, Yabin Chen, Ketao Yin, Jiejuan Yan, Hengzhong Zhang, Zhenxian Liu, John S. Tse, Junqiao Wu, Ho-kwang Mao, and Bin Chen, Large bandgap of pressurized trilayer graphene[J]. PNAS, 2019, 116: 9186 − 9190.
[10] Jianing Xu et al., to be submitted.
[11] Hongliang Dong et al., in preparation.
[12] Jikun Chen et al. Pressure Induced Unstable Electronic States upon Correlated Nickelates Metastable Perovskites as Batch Synthesized via Heterogeneous Nucleation [J]. Advanced Functional Materials, 2020, 30: 23.

A General Mechanism of Grain Growth: Theory and Experimental

Hu Jianfeng*

School of Materials Science and Engineering, Shanghai University

* Corresponding author: jianfenghu@shu.edu.cn

【Introduction】

The practical performances of polycrystalline materials (especially for nano-grained materials) are strongly affected by the formed microstructure inside, which is mostly dominated by grain growth behaviors. The classically quantitative models, such as classical von Neumann-Mullins relation, Hillert's theory, and LSW theory, generally describe normal grain growth (NGG) but fail to interpret quantitatively how abnormal grain growth (AGG) occurs and how grain growth stops. Therefore, it is commonly believed that NGG and AGG are dominated by the different mechanisms of grain boundary (GB) migration. Hence, despite the recent great advances in the understanding grain growth as reviewed in our manuscript, there is still a lack of a quantitative model for AGG, let alone a general model for grain growth behaviors including NGG and AGG. This has puzzled the materials science community for more than 70 years due to its complex nature. Here, we propose a general theory of grain growth and derive a mathematical expression to describe the general grain growth behaviors in polycrystalline systems. The expression depicts how the variables dominate the growth and stagnation of individual grains and thus reveals how AGG and NGG occur in polycrystalline systems. Furthermore, the general growth theory well interprets the counter-intuitive phenomenon of smaller nano-sized grains with higher thermal stability that recently reported in the literature. Meanwhile, we carried out the grain growth experiments to verify the general growth theory. The general evolution of grain growth behaviors including the occurrences of AGG and NGG in the experiments is well interpreted by the general growth theory. Meanwhile, the evolution of grain size distributions of AGG and NGG are in accord with theoretical predictions of the general growth theory. Furthermore, the transition of growth behaviors from AGG to NGG observed in the experiments is consistent with the description of the expression in our model, which corresponds to the GB roughening transition. Therefore, the grain growth experiments confirm the validity of our model.

【Results】

1. Theory

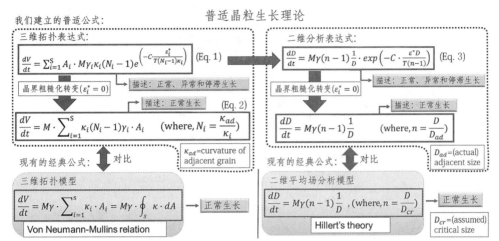

Fig. 1 The general grain growth theory

2. Experimental

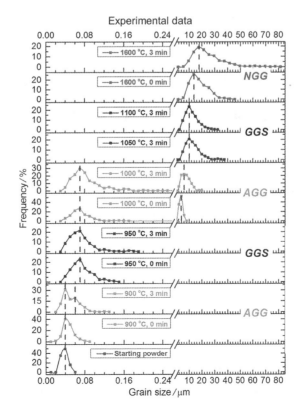

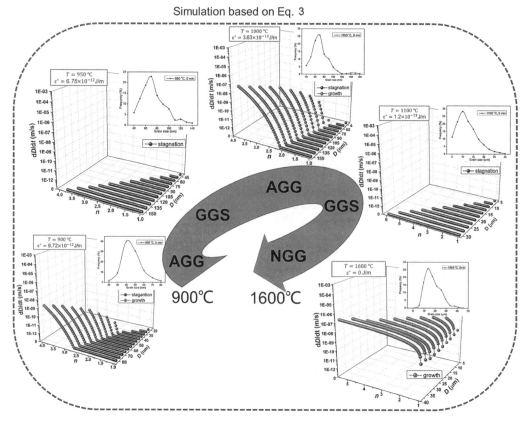

Fig. 2 Experimental results of grain growth

Transition Metal Oxides Induced Spin-Orbit Coupling at Surface Conducting Diamond

Xing Kaijian[1,*], Qi Dongchen[1,2], Christopher I. Pakes[1]

1 Department of Chemistry and Physics, La Trobe Institute for Molecular Science, La Trobe University, Bundoora, Victoria 3086, Australia 2 Centre for Materials Science, School of Chemistry and Physics, Queensland University of Technology, Brisbane, Queensland 4001, Australia

* Corresponding author: 18737250@students.latrobe.edu.au / K.Xing@latrobe.edu.au

【Introduction】

Despite being a bona fide wide-bandgap insulator, hydrogen-terminated diamond, once exposed to moist air, develops an intriguing *p*-type surface conductivity. The underlying mechanism, known as surface transfer doping (STD), induces a spontaneous sub-surface, quasi-two-dimensional (Quai-2D) hole accumulation layer with an areal density of typically 10^{13} cm^{-2} without the need to introduce impurities into the diamond lattice[1,2]. This simple yet effective doping scheme overcomes the limitation caused by conventional bulk doping (e.g. by boron substitution)[3], and also offers a platform for the exploration of quantum transport phenomena in the resulting 2D hole gas (2DHG) at low temperature. As reported, a higher hole density is expected to yield an even stronger SOI[4] and permit an even wider dynamic range of SOI to be achieved. Recently, there have been extensive reports demonstrating those transition metal oxides (TMOs), such as MoO$_3$[5], act as excellent surface acceptors which can enhance the STD efficiency in terms of high hole density comparing with air-doped strategy. However, low-temperature phase-coherent quantum transport in the 2DHG on diamond induced by TMOs transfer doping has so far not been explored. In this work, low-temperature magnetotransport is used as a tool to show the presence of a k^3 Rashba-type SOI with a high spin-orbit coupling of 19.9 meV for MoO$_3$ doping and 22.9 meV for V$_2$O$_5$ doping, respectively, through the observation of a transition in the phase-coherent backscattering transport from weak localization (WL) to weak antilocalization (WAL) at low temperature. It opens up possibilities to study and engineer spin transport in a carbon material system.

【Experimental】

Hall bar devices were fabricated using commercial (Element Six) type-IIa 100 - orientated single-crystal diamond substrates. Hydrogen-termination was carried out via exposure to a microwave hydrogen plasma. Hall bar devices were subsequently fabricated using standard photolithography techniques. Following the fabrication of the diamond Hall bars, the sample was annealed to a temperature of 400℃ for 30 minutes in an ultra-high vacuum (UHV) system with a base pressure of 7×10^{-10} mbar to completely remove adsorbed water and atmospheric adsorbates while keeping the hydrogen-termination intact. In the same vacuum, MoO$_3$ (Device A) and V$_2$O$_5$ (Device B) were thermally deposited via standard Knudsen cells (Fermion Instrument) onto the active region of the Hall bars through a precisely-aligned shadow mask as shown in Fig. 1.

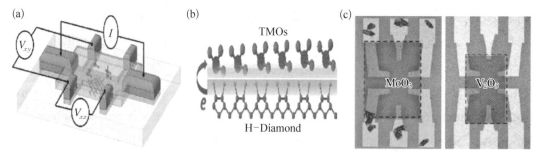

Fig. 1 (a) Configuration of TMO (MoO$_3$, V$_2$O$_5$) doped Hall bar devices; (b) cross-section of the TMO/hydrogen-terminated-diamond interface with electron exchange; (c) Optical image of TMO (MoO$_3$, V$_2$O$_5$) doped Hall bar device

【Results】

WAL arises when a material experiences strong SOI and is typically seen in 2D electronic systems in which the SOI is further enhanced by the reduced dimension. The SOI exists only in crystals without buck or surface inversion symmetry, such as the Dresselhaus effect or Rashba effect. In hydrogen-terminated diamond, the WAL may arise from either anisotropy of the crystal, due to the hydrogen termination and (2 × 1) surface reconstruction, or from the asymmetry of the 2D surface confining potential that leads to the surface hole accumulation. Our previous work in air-induced conducting devices showing that the SOI can be tuned over a wide range with an electrostatic gate suggests that the Rashba effect is the dominant mechanism[4]. We, therefore, use the 2D localization theory, appropriate for a k^3 Rashba interaction in a 2D hole system, to describe the observed changes in magnetoconductivity caused by WL and WAL in the current case:

$$\frac{\Delta\sigma}{G_0} = -\left\{ \psi\left(\frac{1}{2} + \frac{B_\varphi + B_{SO}}{B}\right) + \frac{1}{2}\cdot\psi\left(\frac{1}{2} + \frac{B_\varphi + 2\cdot B_{SO}}{B}\right) \right.$$
$$- \frac{1}{2}\cdot\psi\left(\frac{1}{2} + \frac{B_\varphi}{B}\right) - \ln\left(\frac{B_\varphi + B_{SO}}{B}\right)$$
$$\left. - \frac{1}{2}\cdot\ln\left(\frac{B_\varphi + 2\cdot B_{SO}}{B}\right) + \frac{1}{2}\cdot\ln\left(\frac{B_\varphi}{B}\right) \right\} \quad (1)$$

where ψ is the digamma function. The characteristic phase- and spin-fields are B_φ and B_{SO}, which can be defined by $B_\varphi = \hbar/(4eD\tau_\varphi)$ and $B_{SO} = \hbar/(4eD\tau_{SO})$, respectively. Here, τ_φ is the phase coherence time, τ_{SO} is the spin coherence time. To compare and model the magnetoconductivity data for different temperatures, we plot in Fig. 2(a) and (b) the change in low field magnetoconductivity $\Delta\sigma = \sigma(B) - \sigma(B=0)$ in units of

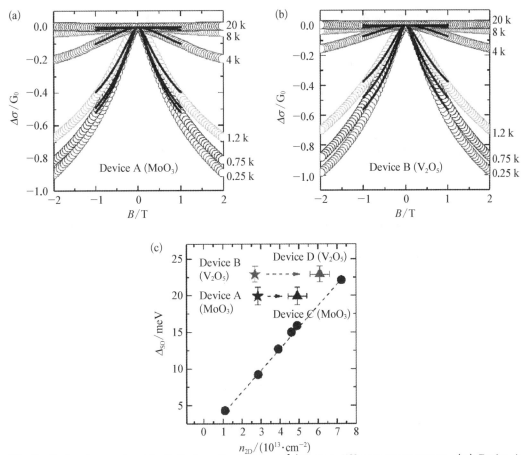

Fig. 2 Reduced magnetoconductivity in units of $G_0 = e^2/\pi h$ at a different temperatures: (a) Device A (MoO$_3$); (b) Device B (V$_2$O$_5$). The open circles represent the experimental data, and the solid lines are fits to theory. (c) Spin-orbit interaction strength (Δ_{SO}) plotted as a function of hole density

$G_0 = e^2/\pi h$. The open circles represent the experimental data at different temperatures. The black lines in Fig.2(a) and (b) correspond to a least-square fit of the experimental data using equation (1) and good agreement with the experimental data is observed at all temperatures within the diffusive regime ($B \ll B_{tr}$). The single theoretical model describes well the transition from WL to WAL as the temperature decreases. Having quantified WL and WAL from the experimental magnetoconductivity curves, we now turn to the evaluation of the spin-orbit interaction strength, Δ_{SO}, which is 19.9 ± 2 meV for Device A (MoO$_3$) and 22.9 ± 2 meV for Device B (V$_2$O$_5$) hydrogen-terminated-diamond interface as shown in Fig.2(c). In both cases, the SOI strength is found to be at least two times higher than the value of 9.74 meV determined for the air-induced surface conducting diamond devices[5]. After hole density correction, two data points fall onto the universal curve reported in our previous work[4], suggesting that the high SOI for TMOs-induced 2D conducting layers is possibly derived from local regions of the surface. (Further explanation is published in *Carbon* DOI: org/10.1016/j.carbon.2020.03.047)

References:

[1] Maier F, Riedel M, Mantel B, et al. Origin of Surface Conductivity in Diamond[J]. Physical Review Letters, 2000, 85(16): 3472-3475.

[2] Pakes C I, Garrido J A, Kawarada H. Diamond Surface Conductivity: Properties, devices, and Sensors[J]. Mrs Bulletiu, 2014, 39(6): 542-548.

[3] Kalish, Rafi. Diamond as a unique high-tech electronic material: difficulties and prospects[J]. Journal of physics D Applied physics, 2007, 40(20): 6467-6478.

[4] G Akhgar, O Klochan, Zaurens H, et al. Strong and Tunable Spin-orbit Coupling in a Two-Dimensional Hole Gas in Ionic-Liquid Gated Diamond Devices. Nano Letters, 2016, 16(6): 3768-3773.

[5] Xing K, Xiang Y, Jiang M, et al. M_0O_3 induces P-type Surface Conductivity by Surface Transfer Dopiry in Diamond[J]. Applied Surface science, 2019, 509: 144890.

The Preparation, Characterization and Application of Nanostructured Mg-based Hydrogen Storage Materials

Zou Jianxin[1,2,*], Lu Chong[2], Zhang Qiuyu[1,2], Ding Wenjiang[1,2]

1 Shanghai Jiao Tong University, National Engineering Research Center of Light Alloys Net Forming & State Key Laboratory of Metal Matrix Composites, Shanghai 200240, China 2 Shanghai Jiao Tong University, Center of Hydrogen Science & School of Materials Science and Engineering, Shanghai 200240, China

* Corresponding author: zoujx@sjtu.edu.cn

【Introduction】

Mg is recognized as one of the most promising hydrogen storage materials due to its low cost and high hydrogen storage density. However, the high desorption temperature and slow desorption kinetics seriously restrict its practical applications. Nanostructuring Mg-based materials is a prospective approach to improve the de/

hydrogenation performances owing to the simultaneous enhancement of the thermodynamic and kinetic properties of Mg. In this work, the nanostructured Mg@TM (TM=Pt, Co, V, Ti) composites with core-shell structure were synthesized *via* a facile method. Moreover, the preparation, characterization, and application of nanostructured Mg-based materials were systematically investigated.

【Experimental】

The nanostructured Mg@TM (TM=Pt, Co, V, Ti) composites were prepared *via* the facile approach containing an arc plasma method followed by the electroless plating in solutions. In particular, the $TMCl_x$ solutions with adding pure Mg ultrafine powder with a fixed molar ratio were stirred vigorously for a certain duration at room temperature. Then the resultant products were isolated through centrifugation and rinsed repeatedly to wash away the waste (residual $MgCl_2$). Finally, the nanostructured Mg@TM composites were obtained by drying the rinsed products under vacuum.

【Results】

Core-shell structured Mg@TM (TM=Pt, Co, V, Ti) composites were obtained *via* the approach containing an arc plasma method followed by electroless plating in solutions. The phase compositions, microstructures, and hydrogen storage properties of the nanoscale Mg@TM (TM = Pt, Co, V, Ti) composites were carefully investigated. The Mg@TM composites consisted of a 50 – 200 nm Mg core and an approximately 10 nm TM shell. All of the Mg@TM (TM=Pt, Co, V, Ti) composites showed the improved hydrogen storage performances in comparison with the pure Mg powders. The hydrogenation activation energies of Mg@TM (TM=Pt, Co, V, Ti) composites reduced to 82.4, 73.2, 86.3, and 72.2 kJ mol^{-1} H_2, respectively. Moreover, the onset dehydrogenation temperatures of the hydrogenated composites were 340, 346, 342, and 367 ℃, which were much lower than that of pure MgH_2 (383℃). XRD and HRTEM results revealed the existence of Mg_3Pt, Mg_2CoH_5, VH_2, and TiH_2 in the hydrogenated Mg@TM (TM=Pt, Co, V, Ti) composites, acting as "hydrogen pump" to accelerate the hydrogen sorption. Besides, the core-shell structures could introduce more active interfaces and nucleation sites for hydrogen de/sorption, which is also contributed to the improved hydrogen storage properties of Mg. The nanostructured Mg-based materials were employed as the hydrogen-provider to the fuel cell for the unmanned aerial vehicle and large scale hydrogen transportation.

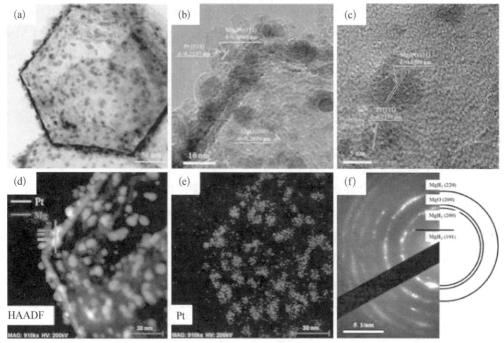

Fig. 1　TEM and HADDF-STEM observations of the hydrogenated Mg@Pt composite

Design of Sulfur Cathode Host Material Based on Strong Polarity

Liu Handing, Man Han, Pei Ke, Fang Fang*, Che Renchao*, Sun Dalin*

Fudan University

* Corresponding author: f_fang@fudan.edu.cn

【Introduction】

Perceived as a promising candidate for secondary ion battery, lithium-sulfur has been a critical device for energy storage and conversion. Sulfur cathode exhibits remarkable properties such as high theoretical specific capacity and power density, low cost, and high natural abundance, making a great match with lithium metal anode. However, the poor intrinsic conductivity of sulfur has led to expectations of developing host material with higher conductivity to promote electrons transfer. Because of the weak polarity of the host material, the inferior combination with the intermediate lithium polysulfide is prone to cause the loss of sulfur cathode during the long cycle, rendering serious degradation of the battery performance. Therefore, investigations of the host material with high polarity is revealed as an essential part of boosting the comprehensive electrochemical performance and further commercializing of the sulfur cathode.

【Experimental】

The composite system of metal oxide and carbon matrix was synthesized by the combination of simple hydrothermal reaction, high-temperature carbonization, and melting impregnation. First, Fe_2O_3 nanoparticles, applied as a soft template, were coated with a nitrogen-doped mesoporous carbon layer and etched into nanoparticles with smaller size and better dispersivity. Afterward, ultra-thin Mn_3O_4 nanosheets were grown on the surface of the mesoporous carbon shell via solvothermal treatment. Finally, after a sulfur infiltration treatment, the composite sulfur cathode (Fe_2O_3@N-PC/Mn_3O_4-S) was obtained. By comparison of electrochemical performances of the sulfur cathode with various host material, the impact of the polarity of host material on electrochemical performance was analyzed. The essence of the polarity variation of the host material was also discussed by combination analysis of in-situ electron microscopy and electron holography.

【Results】

Rambutan-like sulfur cathode with a yolk-shell structure was prepared by coating Fe_2O_3 nanoparticles with a nitrogen-doped mesoporous carbon layer modified with Mn_3O_4 nanosheets (Fe_2O_3@N-PC/Mn_3O_4). When applied as a cathode for Li-S battery, the as-prepared cathode delivered an initial discharge capacity of 1 425 mAh·g^{-1} at a current density of 0.5 C and remained a capacity of 1 122 mAh·g^{-1} after 400 cycles. At a high current density of 10 C, the capacity stabilized at 639 mAh·g^{-1} after 1 500 cycles. Based on the analysis, Fe_2O_3 nanoparticles and ultra-thin Mn_3O_4 nanosheets from the internal and external carbon shels in the Fe_2O_3@N-PC/Mn_3O_4 structure could generate a coupling effect, increasing the integral oxygen vacancy concentration of the host material. This coupling effect not only improves the overall conductivity of the cathode material but also enhances the polarity of the composite system, leading to a stronger anchoring effect for lithium polysulfide. From the above discussion, Fe_2O_3@N-PC/Mn_3O_4-S cathode exhibits ultra-stable cycling performance and outstanding rate capacity.

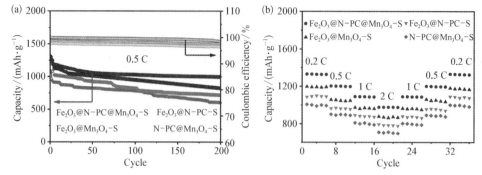

Fig. 1 (a) Long-cycling performance at a current density of 0.5 C, (b) rate capabilities of Fe_2O_3@N-PC/Mn_3O_4-S, Fe_2O_3@Mn_3O_4-S, Fe_2O_3@N-PC-S, and N-PC/Mn_3O_4-S cathodes

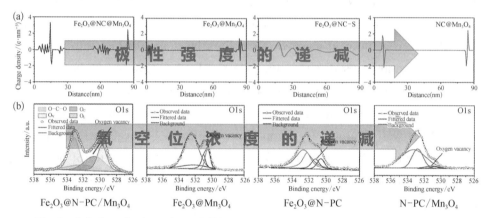

Fig. 2 (a) the polar intensity and (b) oxygen vacancy concentration contrast of Fe_2O_3@N-PC/Mn_3O_4, Fe_2O_3@Mn_3O_4, Fe_2O_3@N-PC, and N-PC/Mn_3O_4 host

A Facile Applicable Strategy for Construction of 3D Porous Gelatin-Alginate Hydrogels for Deep Second-Degree Scald Wound Healing

Xiao Xu, Lin Che, Ling Xu*

State Key Laboratory of Molecular Vaccinology and Molecular Diagnostics, School of Public Health, Xiamen University, Xiamen 361102, China
* Corresponding author: lingxu@xmu.edu.cn

【Introduction】

Globally, burns are a serious public health problem. Annually, about 300 000 people die or suffer injuries caused by fires, scalds, electrical burns, and other forms of burns. Moreover, burns from scalds and fires account for approximately 80% of all reported burns. Scalds have some distinct pathophysiological characters among all types of burns, such as more wound exudate and necrotic tissue debris on the surface of the burn wounds. The burn wound is classified into different degrees dependent upon the thickness of burn injuries. The lower layers of the dermis were severely damaged in the deep second-degree burns. The deep second-degree burns wound healing is a complex physiological process in which skin tissue is repaired in the case of injury. During wound healing, the wound area is covered with necrotic tissue, and this inactive skin of necrotic tissue often hinders wound repair.

【Experimental】

The introduction of electron beam (EB) radiation crosslinking technology to both alginate, gelatin, and CMC allows the formation of a biodegradable three-dimensional (3D) porous network upon applying EB-irradiation, which is a straightforward, fast, and cost-effective approach. This study focused on the synthesis and comprehensive

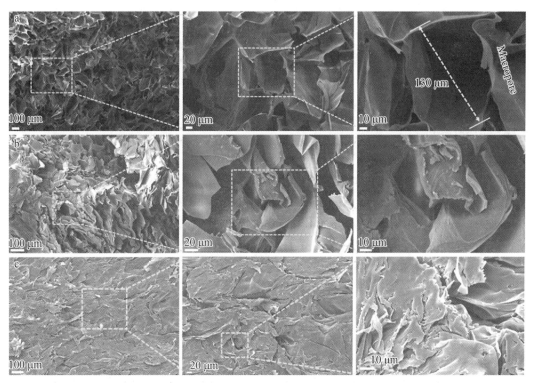

Fig. 1 Surface morphology characterization for gelatin-alginate hydrogels. SEM images of the (a) GSC222, (b) GSC213, and (c) GSC231 gelatin-alginate hydrogels at different magnifications

characterization (FTIR, TGA, SEM) of environmentally friendly hydrogels based on gelatin, CMC, and alginate for burn wound repair substitutes.

【Results】

We present a facile applicable strategy for crosslinking gelatin-alginate hydrogels into a biodegradable three-dimensional (3D) porous. In vivo results show that the deep second-degree scald wound-healing process is obviously accelerated.

Microstructures and Optoelectronic Properties of NiO Films Deposited by High Power Impulse Magnetron Sputtering

Sun Hui[1,*], Chen Shengchi[2], Kuo Tsungyen[2], Wang Kunlun[1], Song Shumei[1]

1 School of Space Science and Physics, Shandong University, Weihai, Shandong, 264209, China   2 Department of Materials Engineering and Center for Plasma and Thin Film Technologies, Ming Chi University of Technology, Taipei 243, Taiwan, China

*Corresponding author: huisun@sdu.edu.cn

【Introduction】

Within a few intrinsic p-type transparent conductive oxides (TCOs), NiO with wide bandgap is a promising candidate. Due to its resource availability, low production cost, and non-toxicity, NiO film can be employed in various fields. In order to clarify the p-type conductivity mechanism of NiO, a series of work has been reported. It is reported that the p-type conductivity of NiO maybe derives from the conversion of Ni^{2+} to Ni^{3+} under oxygen-rich conditions. In this process, Ni vacancies and holes are associated, resulting in an increment in carrier concentration and film's conductivity. Recently, the high power impulse magnetron sputtering (HiPIMS) technology developed on the basis of conventional magnetron sputtering has attracted people's attention because of its extremely high target ionization rate. The high target ionization rate of this technology is beneficial to increase Ni^{3+} content level during the film's formation, thereby improving the probability of the generation of Ni vacancies, which in turn enhances the film's p-type conductivity.

【Experimental】

In the current work, NiO films were deposited through HiPIMS technology. The peak power density supplied to the Ni target is adjusted by changing the duty cycle, thereby affecting the target ionization rate. The contribution of enhanced ionization rate to the optoelectronic properties of p-type NiO films was investigated.

【Results】

The analysis of the chemical bonding state in NiO films by X-ray photoelectron spectroscopy confirms the variation of Ni^{3+} content with pulse off-time rising. The curve fitting of $Ni2p^{3/2}$ spectrum of NiO films deposited with pulse off-time of 0, 1 000, 2 000, and 3 000 μs is compared. Ni^{3+} ions are detected in both films, revealing that holes are generated in both films, consistent with the results of p-type conductivity for all films. Moreover, upon extending pulse off-time, the Ni^{3+} integral area is greatly increased. The atomic ratio of Ni^{3+}/Ni^{2+} increases from 1.8 to 3.6 as pulse off-time varies from 0 to 3 000 μs, indicating the holes' density related to Ni^{3+} increases considerably. This result proves that HiPIMS technology is beneficial to enhance the carrier concentration in NiO films and thereby improve the film's p-type conductivity.

HiPIMS technology with a high ionization rate is confirmed as a preferred technology for preparing NiO film possessing a high Ni^{3+}/Ni^{2+} ratio which contributes to the film's p-type conductivity. With increasing pulse off-time, the film's orientation changes from (111) to (200) plane. Compared to the film deposited in DC sputtering mode, the film's resistivity significantly reduces for the films produced by HiPIMS. This is mainly caused by the elevation of carrier concentration, which results from the high density of ionized Ni^{3+} produced under longer pulse off-time conditions. In contrast, the film's transmittance degrades with prolonging pulse off-time. It is caused by the refinement of the grain size and the enhancement of Ni vacancy density resulting in the light scattering and absorption. Our endeavors prove that HiPIMS is a preferred technology for preparing NiO films with high p-type conductivity.

Pressure-induced Dramatic Changes in Halide Perovskites

Lyu Xujie*, Guo Songhao, Luo Hui, Wang Yingqi, Yang Wenge

Center for High Pressure Science and Technology Advanced Research

*Corresponding author: xujie.lu@hpstar.ac.cn

【Introduction】

Metal halide perovskites have emerged as a promising class of materials for advanced optoelectronic applications. Solar cells and light-emitting diodes based on halide perovskites have achieved an impressive power conversion efficiency of 25.2% and a high external quantum efficiency of over 20%, respectively. Moreover, exceptional structural turnability enables enhanced and/or emergent properties. By using appropriate organic and inorganic components, various perovskite-like materials ranging from 3D to 2D, and 1D structures on the molecular level can be obtained. However, challenges remain including the low stability and the lack of an insightful understanding of the structure-property relationships.

【Experimental】

Diamond anvil cells (DACs) were employed to generate high pressure. Samples and ruby microspheres (for pressure measurements) were loaded in the DAC chamber. *In situ* high-pressure characterization methods including X-ray diffraction, Raman, photoluminescence, UV-Vis spectroscopy, electrical conductivity, photocurrent measurements were carried out. First-principles calculations were also performed to help the understanding of the underlying mechanisms.

【Results】

In our studies, we have used an alternative means, pressure, to tune the structures and physical properties of halide perovskites with different dimensionality. Using state-of-the-art high-pressure techniques coupled with *in situ* synchrotron-based and

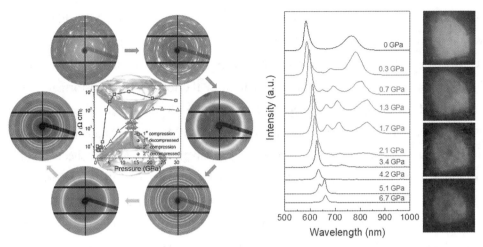

Fig. 1　Pressure-induced variations of structure and properties of halide perovskites

in-laboratory property measurements, we characterized the changes in their structural, electrical, and optical properties. Pressure-enhanced properties such as higher electron transport and stronger light absorption, as well as pressure-induced novel phenomena including emergent photocurrent, were observed. Our findings reveal that high pressure can potentially be able to realize enhanced and/or emergent properties of the halide perovskites, and further our understanding of the fundamental structure-property relationships.

Topological Transformation of Layered Double Hydroxide Nanosheets for Efficient Photocatalytic CO_2 reduction

Zhao Yufei[1,*], Song Yufei[1], Duan Haohong[2]

1 State Key Laboratory of Chemical Resource Engineering, Beijing University of Chemical Technology　2 Tsinghua University

* Corresponding author: zhaoyufei@mail.buct.edu.cn

【Introduction】

Layered double hydroxide (LDHs) represent an important class of 2D layered materials with the general formula of $[M^{2+}_{1-x}M^{3+}_x(OH)_2]^{x+}(A^{n-})_{x/n} \cdot mH_2O$[1]. Compared with other oxides (hydroxides), LDHs show the advantage of tunable chemical composition, thickness, thereby can be widely used as catalysts[2]. The topological transformation of layered double hydroxides (LDHs) to the corresponding mixed metal oxides (MMO) with highly exposed active facets offers an efficient strategy to achieve such interfaces. However, the formation and conversion of these heterostructured interfaces is the lack of study and remains to be elusive. Herein, we report a detailed investigation of the topological transformation of Zn/Ni-Co/Ti-LDHs by using X-ray absorption fine structure (XAFS) measurements and density functional theory (DFT) calculation. The as-prepared MMO with abundant interfaces and exposed highly active facets can be modulated by calcination of LDH nanosheets at different temperatures. Especially, the ZnCo-LDH calcinated at 200℃ gives the formation of high-energy {112} faceted spinel-Co_3O_4 nanosheets confined by highly active {001} faceted ZnO nanosheets with abundant interfaces and exhibits excellent performance in electrochemical water oxidation and photothermal CH_4 activation[3-5]. Through adjusting the composite of LDH, {112} faceted NiO and {001} faceted TiO_2 nanosheets can be obtained. This work will deepen the understanding of the LDH topological transformation from the molecular level, thus providing a strategy for the controllable synthesis of high-performance catalysts through the topological transformation of LDHs.

【Experimental】

The ZnCo-LDH and CoCo-LDH were prepared by the co-precipitation method. The corresponding calcined derivatives were denoted as ZnCo-x and CoCo-x (x refers to the different calcination temperatures at 200, 400, 600, and 800 ℃). Owing to the abundant interfaces, ZnCo-200 possesses the smallest bandgap and exhibits better OER performance than others calcinated at higher temperatures (Fig. 1).

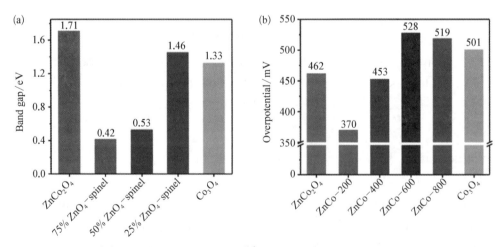

Fig. 1 (a) Band gaps of $ZnCo_2O_4$, $y\%$ ZnO_4-spinel, and Co_3O_4, respectively;
(b) Overpotential of $ZnCo_2O_4$, ZnCo-x, and Co_3O_4, respectively

【Results】

During the calcination from room temperature to 200 ℃, the collapse of the LDH structure occurs with the co-formation of (112)-faceted spinel-like Co_3O_4 and (001) exposed zincite-structured ZnO nuclei with abundant interfaces. Especially, the interfaces can be identified to the Zn^{2+} ions doped in tetrahedral sites (T_d) into the spine-like Co_3O_4 as $ZnCo_2O_4$, and Co-doped ZnO phase with a slight doped content (<5%), as evidenced by some of our previous work. Further increase of the calcination temperature from 200 to 400 ℃, the crystallinity and size of both spine-like Co_3O_4 and zincite-structured ZnO increase, and the boundaries between these two phases become less. Further increasing the calcination temperature to 600 ℃, the doped Zn^{2+} ions are gradually leaching from the spinel-like Co_3O_4. The reason may be explained to the cation migration at high temperatures[5]. It should be noted that the interface boundaries become clearly decreasing. At 800 ℃, the size of the two corresponding phases (spinel-like Co_3O_4 and zincite-structured ZnO) sheets become bigger and the interfaces almost disappear because of the sintering. At this time, the Zn^{2+} ions are mostly removed from the spinel-like Co_3O_4 and the phase separation is obvious(Fig. 2).

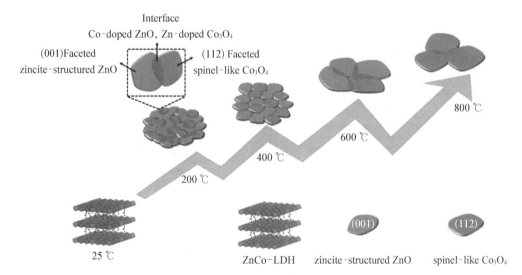

Fig. 2 Schematic illustration of the topological transformation from the ZnCo-LDH to the ZnCo-x at different temperatures

References:

[1] Zhao, Y., G.I.N. Waterhouse, G. Chen, X. Xiong, L.Z. Wu, C.H. Tung, and T. Zhang, Two-dimensional-related catalytic materials for solar-driven conversion of CO_x into valuable chemical

feedstocks[J], Chem. Soc. Rev, 2019, 48(7): 1972-2010.

[2] Tan, L., S.M. Xu, Z. Wang, Y. Xu, X. Wang, X. Hao, S. Bai, C. Ning, Y. Wang, W. Zhang, Y.K. Jo, S.J. Hwang, X. Cao, X. Zheng, H. Yan, Y. Zhao, H. Duan and Y.F. Song, Highly selective photoreduction of CO_2 with suppressing H_2 evolution over monolayer layered double hydroxide under irradiation above 600 nm[J], Angew. Chem. Int. Ed, 2019, 58(34): 11860-11867.

[3] Xu, Y., Z. Wang, L. Tan, H. Yan, Y. Zhao, H. Duan and Y.-F. Song, Interface Engineering of High-Energy Faceted Co_3O_4/ZnO Heterostructured Catalysts Derived from Layered Double Hydroxide Nanosheets[J], Ind. Eng. Chem. Res, 2018, 57(15): 5259-5267.

[4] Xu, Y., Z. Wang, L. Tan, Y. Zhao, H. Duan and Y.-F. Song, Fine Tuning the Heterostructured Interfaces by Topological Transformation of Layered Double Hydroxide Nanosheets[J], Ind. Eng. Chem. Res, 2018, 57(31): 10411-10420.

[5] Zhao, Y., X. Jia, G. Chen, L. Shang, G.I. Waterhouse, L.Z. Wu, C.H. Tung, D. O'Hare, and T. Zhang, Ultrafine NiO Nanosheets Stabilized by TiO_2 from Monolayer NiTi-LDH Precursors: An Active Water Oxidation Electrocatalyst[J], J. Am. Soc. Chem., 2016, 138(20): 6517-6524.

High-efficiency CO_2 Separation via Reaction-promoted Transport in Two-dimensional Sub-nanometer Channels

Xu Xiaozhi, Shi Kaiqiang, Li Biao, Han Jingbin*

Beijing University of Chemical Technology
* Corresponding author: hanjb@mail.buct.edu.cn

【Introduction】

Membrane-based gas separation is a promising separation method, but the development of membrane materials with simultaneously high selectivity and permeability remains a major challenge for practical application. Hybrid membranes with two-dimensional sub-nanometer channels were fabricated by self-assembly of unilamellar layered double hydroxide (LDH) nanosheets and formamidine sulfinate (FAS), followed by spray-coating of a poly(dimethylsiloxane) (PDMS) layer. High-efficiency CO_2 separation performance was realized based on the combination of enhanced solubility, diffusivity, and reactivity in the interlayer space. The CO_2 solution selectivity was achieved by introducing the CO_2-philic LDH nanosheets, while a suitable interlayer spacing of 0.34 nm allowed the permeation of CO_2 with the rejection of larger-sized gas molecules. It is of utmost importance that the FAS between the neighboring LDH nanosheet acts as a CO_2 carrier facilitates the permeation of CO_2 via reversible chemical reactions. Thus, the as-prepared (LDH/FAS)$_n$-PDMS membrane exhibits excellent CO_2 preferential permeability (CO_2 permeability: 162.7 Barrer, CO_2/N_2, CO_2/H_2, and CO_2/CH_4 selectivity: 84.96, 43.04 and 62.34 respectively), breaking through the selectivity-permeability trade-off dilemma in CO_2 separation.

【Experimental】

As-prepared (LDH/FAS)$_n$-PDMS with ordered superlattice structure were fabricated *via* layer-by-layer (LBL) assembly of MgAl-LDH nanosheets and formamidine sulfinic acid (FAS) alternatively, followed by coating a thin layer of poly (dimethylsiloxane) (PDMS). The preparation process and structure of (LDH/FAS)$_n$-PDMS membrane were characterized with atomic force microscope (AFM), X-ray diffraction (XRD) spectrum, particle size distribution, scanning electron microscope (SEM), high-resolution transmission electron microscope (HRTEM), UV-Vis absorption spectra, FT-IR spectrum, *in situ* diffuse reflectance infrared Fourier-transform spectroscopy (DRIFTS), X-ray photoelectron spectroscopy (XPS) spectra, thermogravimetric analysis (TGA), water contact angles, CO_2 temperature-programmed desorption (TPD) profile and CO_2/N_2 adsorption isotherm. CO_2 separation performance was measured using a VAC-V2 gas transmission rate testing system.

【Results】

Upon increasing n from 5 to 25, the (LDH/FAS)$_n$-PDMS membrane exhibits an enhanced ideal selectivity of CO_2/H_2, CO_2/N_2, and CO_2/CH_4 from 0.60, 0.90, and 0.78 to 43.04, 85.63, and 62.34, respectively. Compared with these upper bound lines and other membrane materials in the literature, the CO_2 permselectivity of (LDH/FAS)$_{25}$-PDMS membrane outperforms most of the reported values and is higher than the Robeson or Freeman upper bound limits. These results reveal that the (LDH/FAS)$_{25}$-PDMS membrane has overcome the "trade-off" effect between permeability and selectivity, providing an efficient CO_2 separation material for industrial gas mixtures with the practically applied foreground. The (LDH/FAS)$_n$-PDMS membranes exhibit excellent CO_2 permselectivity as well in the mixed gas system. [As shown in

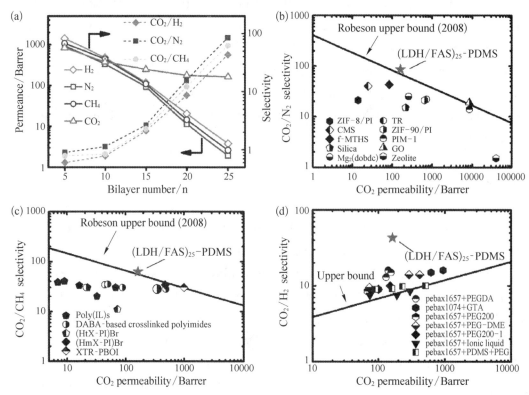

Fig. 1 (a) The H_2, CO_2, N_2, CH_4 transmission rates and CO_2/H_2, CO_2/N_2, CO_2/CH_4 selectivity for $(LDH/FAS)_n$-PDMS membrane; (b) CO_2/N_2; (c) CO_2/CH_4; and (d) CO_2/H_2 separation performance of $(LDH/FAS)_{25}$-PDMS membrane and representative membranes reported in the literature

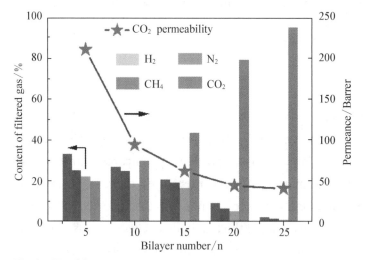

Fig. 2 The CO_2 permeability and H_2, CO_2, N_2, CH_4 contents in the filtered gas for $(LDH/FAS)_{25}$-PDMS membrane under mixed gas feed condition

Fig.4, when the bilayer number is small ($n=5$), the contents of H_2, N_2, CH_4, and CO_2 in the filtered gas are 33.1%, 22%, 25.2%, and 19.7% respectively.] The sub-nanometer channels between LDH and FAS play the role of gas sieving by molecular size; while the hydroxyl groups in the LDH layer increase the affinity of CO_2, leading to improved solubility. Most significantly, the amidine groups in FAS can react with CO_2 selectively and reversibly, promoting the transport of CO_2. With the advantages of precise control on gallery height, order distribution of carriers, and ease of preparation, the lamellar membrane reported here holds great potential in CO_2 capture and separation.

锌离子电池与储能变色双功能器件

康利涛[1,2,*]，崔芒伟[2]，王变[2]

1 烟台大学环境材料学院　2 太原理工大学材料学院

* 通讯作者：kangltxy@163.com

【引言】

锌离子电池（ZIBs）使用金属锌负极和近中性水系电解液，具有安全性好、成本低、污染小、能量密度高等特点。但 ZIBs 的金属锌负极在充放电循环中，会因为"尖端效应"而产生自发、持续的枝晶生长，引起电池内阻增大、容量衰减、甚至刺穿隔膜引起电池短路失效。此外，ZIBs 正极材料种类多样，通过合理电化学路径设计，有望开发出多功能化的 ZIBs 新型器件。本次报告将介绍本课题组在电极及电解液改性提升锌负极循环寿命和储能变色双功能透明锌离子电池的相关研究工作。

【实验】

本报告包含的实验主要有：（1）采用浆料刮涂法，在锌负极表面涂覆纳米 $CaCO_3$ 的多孔绝缘层；（2）采用 SEM 测试常用的喷金工艺，在锌负极表面沉积纳米金颗粒；（3）采用水热法，在 FTO 玻璃表面沉积普鲁士蓝涂层。

【结果】

1. 利用纳米孔道对电解液迁移的均化作用、对锌沉积位点的引导作用、以及绝缘涂

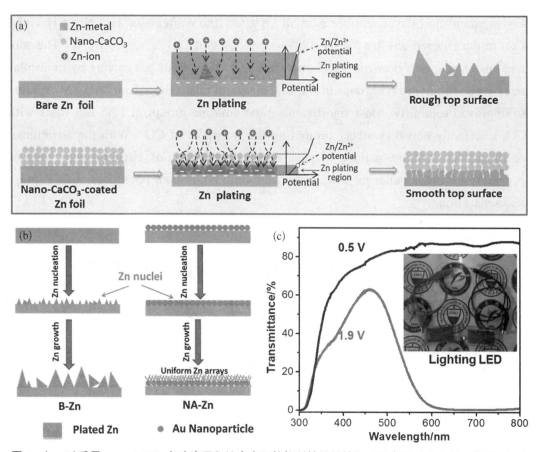

图 1 （a—b）采用 nano-CaCO₃ 多孔涂层和纳米金颗粒抑制锌枝晶的机理示意图；（c）采用双离子电解液制备的 PB/Zn 双功能透明锌电池的透光率曲线与放电功能展示

层对锌沉积反应的限域作用三重机制，实现了自下而上的、均匀稳定的锌沉积-溶解反应，有效提升 ZIBs 的循环稳定性（*AEM*，2018，8，1801090）；

2. 溅射沉积纳米金颗粒可以作为锌沉积的异相成核位点，引导锌均匀沉积，避免大尺寸锌枝晶生长，有效提升 ZIB 循环稳定性（*ACS AEM*，2019，2，6490）；

3. 在透明导电 FTO 玻璃涂层表面沉积 PB 涂层，并使用 K$^+$/Zn^{2+} 双离子电解液串联 PB/PW-K$^+$（普鲁士蓝-普鲁士白）和 Zn/Zn^{2+} 电对，开发可变色（蓝色-透明）透明锌电池，变色时间＜5 s，循环寿命 7 000 圈（*Solar RRL*，2020，4，1900425，Back Cover）。

Explore Potential Advantages of Na-Ion Batteries with Ultralow-Concentration Electrolyte

Li Yuqi[1,2], Lu Yaxiang[1,2], Hu Yongsheng[1,2,*]

1 Institute of Physics, Chinese Academy of Sciences
2 University of Chinese Academy of Sciences
* Corresponding author: ysh@iphy.ac.cn

【Introduction】

Concentration regulation is a key design consideration of electrolytes for advanced energy storage devices. In the past few years, increasing the salt concentration has provided a satisfactory solution for Li-metal batteries, aqueous batteries, etc. On the opposite end, decreasing the salt concentration to form a super dilute electrolyte has been fully unexplored. Owing to the smaller Stokes radius and desolvation energy of Na^+ compared to those of Li^+, an unusual ultralow-concentration electrolyte was proposed for Na-ion batteries (NIBs) to further reduce the cost and expand the working temperature range.

【Experimental】

A class of the Na-based electrolytes with different concentrations was prepared by dissolving $NaPF_6$ in ethyl carbonate (EC) and propylene carbonate (PC). The Na-ion full-cells were assembled with the disordered carbon anode materials and layered O_3-$Na[Cu_{1/9}Ni_{2/9}Fe_{1/3}Mn_{1/3}]O_2$ cathode materials, using the electrolytes with different concentrations. Surface passivation chemistry after one cycle of NIBs from X-ray photoelectron spectroscopy (XPS) experiments with Ar^+ etching.

【Results】

As shown in Fig. 1, it was surprisingly found that NIBs can work well in such ultralow-concentration electrolyte (0.3 M) and demonstrate unexpected advantages. The dilute concentration not only reduces the cost significantly but also expands the working temperature range (-30 to 55 ℃) for durable NIBs. Low viscosity and less corrosive risk (less HF attack) would help improve the interfacial wettability and Columbic efficiency at low and high temperatures, respectively. Moreover, the formed stable organic-dominated solid electrolyte interphase (SEI) or cathode electrolyte interphase (CEI) with superior kinetics enables the stable operation of NIBs at extreme temperatures. The new dilute electrolyte chemistry is expected to extend to other electrolyte systems used in low-cost rechargeable batteries. Effective additives can further improve the performance of the ultralow-concentration electrolyte from bulk and interface regulation in the future.

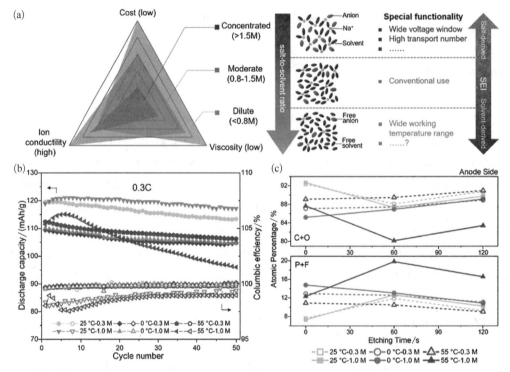

Fig. 1 (a) Overview of the electrolytes with different concentrations. (b) Cyclic capability at 0.3C (30 mA/g) of NIBs. (c) Atomic ratios of C (C 1s) plus O (O 1s) and P (P 2p) plus F (F 1s) elements detected at the SEI films from XPS tests

References:

[1] Yuqi Li, Yaxiang Lu, Philipp Adelhelm, Maria-Magdalena Titirici and Yong-Sheng Hu. Intercalation chemistry of graphite: alkali metal ions and beyond[J]. Chemical Society Reviews, 2019, 48(17): 4655-4687.

[2] Yuqi Li, Yang Yang, Yaxiang Lu, Quan Zhou, Xingguo Qi, Qingshi Meng, Xiaohui Rong, Liquan Chen, and Yong-Sheng Hu. Ultralow-Concentration Electrolyte for Na-Ion Batteries[J]. ACS Energy Letters, 2020, 5(4): 1156-1158.

Study of TiO$_2$ Coated α-Fe$_2$O$_3$ Composites and the Oxygen-defects Effect on the Application as Anode of High-performance Li-ion Battery

Ma Yangzhou*, Zhang Li, Cai Zhenfei, Song Guangsheng*

Key Laboratory of Green Fabrication and Surface Technology of Advanced Metal Materials, Ministry of Education, School of Materials Science and Engineering, Anhui University of Technology, Maanshan, 243000, China
* Corresponding authors: yangzhou.ma@outlook.com; song_ahut@163.com

【Introduction】

α-Fe$_2$O$_3$ is considered to be an excellent candidate as the anode material of the next generation LIBs, which is not only due to its high theoretical capacity (~1 000 mAh·g^{-1}) but also for its low cost, easy preparation, and non-toxicity. Currently, the key issue of the α-Fe$_2$O$_3$ application in LIBs is its low conductivity and electrode pulverization from tremendous volume variation, which results in the low initial coulomb efficiency and rapid capacity decay. It has been demonstrated that the introduction of TiO$_2$ into Fe$_2$O$_3$ is a benefit to improve the cycling stability, however, it is limited by poor electrical conductivity and low chemical diffusivity of lithium for large-scale application. In the present work, a facile sol-gel route was adopted to synthesize TiO$_x$ modified commercial α-Fe$_2$O$_3$ compounds, and it was proved for the first time that the TiO$_2$ with oxygen defects was helpful to improve the α-Fe$_2$O$_3$ cycling stability.

【Experimental】

The TiO$_x$/Fe$_2$O$_3$ composites were synthesized through a facile sol-gel process. Firstly, commercial nano Fe$_2$O$_3$ added into a mixture of deionized water and glycol, and then four different amounts of tetrabutyl titanate (TBOT) were added into the solvent. After stringing for a while, citric acid was added into each mixture. The gel was obtained with continuous mixing and heating after adjusting the pH value to 6. Secondly, the dry gel was heated at 500 ℃ for 3 h and the obtained powder was grinded with a moderate amount of reducing agent for the oxygen defects formation.

【Results】

TiO_2 was well coated on the surface of Fe_2O_3 with an average thickness of 5.5 nm, and the oxygen defects were successfully introduced into the composites with the reduction treatment. Electrochemical characterization showed that TiO_2 coating can improve the cycle performance of Fe_2O_3. TiO_2 effectively prevented the occurrence of irreversible reactions in the first discharge process of Fe_2O_3 anode. The coating layer significantly improved the electronic conductivity and cycling stability of Fe_2O_3 anode material. Moreover, the introduction of oxygen defects possessed more excellent cycling stability than the samples without reduction. The reduced $Fe_2O_3@0.2TiO_2$ sample remained a specific discharge capacity of 405.6 mAh·g^{-1} after 150 cycles, which effectively improved the intrinsic cycling performance of Fe_2O_3, the

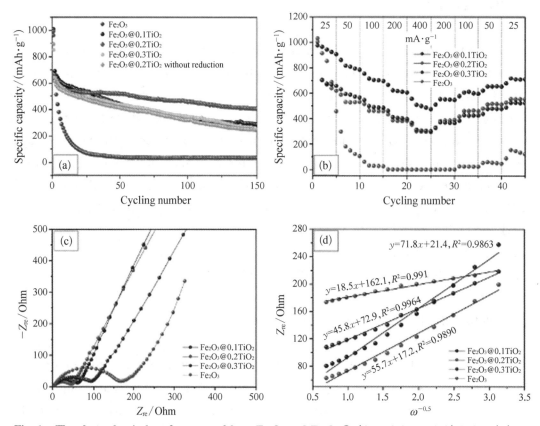

Fig. 1 The electrochemical performance of bare Fe_2O_3 and $Fe_2O_3@(0.1, 0.2$ and $0.3)TiO_2$: (a) the cycling performance at a current density of 100 mA·g^{-1}, (b) the rate capability, (c) Nyquist spectra at the high frequency of electrochemical impedance spectra (EIS), and (d) relationships between Z_{re} and $\omega^{-1/2}$ at low frequency

corresponding discharge capacity of 50 mAh·g^{-1} after 30 cycles.

Design of Advanced Porous Materials for Sodium-ion Batteries

Wang Wanlin[1], Gu Qinfen[2,*], Chou Shulei[1]

1 Institute for Superconducting and Electronic Materials University of Wollongong, Innovation Campus, Squires Way, North Wollongong, NSW 2522, Australia   2 Australian Synchrotron, ANSTO, 800 Blackburn Road, Clayton, Victoria 3168, Australia
* Corresponding author: qinfeng@ansto.gov.au

【Introduction】

Owing to the low cost and natural abundance of sodium, sodium-ion batteries (SIBs) are considered as an alternative to lithium-ion batteries, which is expected to be utilized for gird-scale energy storage in the future[1,2,3]. Different cathode materials for SIBs include a layered oxide[4,5], polyanionic compound[6,7] and Prussian blue analogues (PBAs)[8,9], among them, PBAs have attracted tremendous attention because their open framework structure could easily accommodate Na$^+$ and enable its fast transportation. Compared to other cathodes, the high-temperature calcination is not required during the synthesis of PBAs, which effectively lower the manufacturing costs[8]. These advantages make PBAs quite likely to be mass-produced and widely used as low-cost cathode material for SIBs in the future. Fe-based Prussian blue analogues are considered as low-cost and easily-prepared cathode materials for sodium-ion batteries, their quality, and electrochemical performance are strongly determined by its precipitation process.

【Experimental】

$Na_{2-x}FeFe(CN)_6$ samples were synthesized modified co-precipitation methods at 25℃. The powder and batteries were also examined in situ with the PXRD method on the powder diffraction beamline at the Australian Synchrotron. The obtained powder data from the synchrotron were indexed and refined with TOPAS 5 (Bruker) software. The Morphologies of the as-prepared samples were observed with a field emission scanning electron microscope (SEM, JEOL JSM-7500FA). The element distribution was detected by Energy Dispersive Spectrometer (EDS, JEOL JSM-6490), and more information was obtained with scanning transmission electron microscopy (STEM, JEM-ARM 200F), equipped with selected area electron diffraction pattern. A Mettler-Toledo thermogravimetric analysis/differential scanning calorimetry (TGA/DSC) STARe system was used to determine the water in the sample with a program running from 50 to 500 ℃ ramped at 10 ℃ min^{-1} in Ar. X-ray photoelectron spectroscopy (XPS, PHI5600, PerkinElmer) measurements were performed to obtain the valence

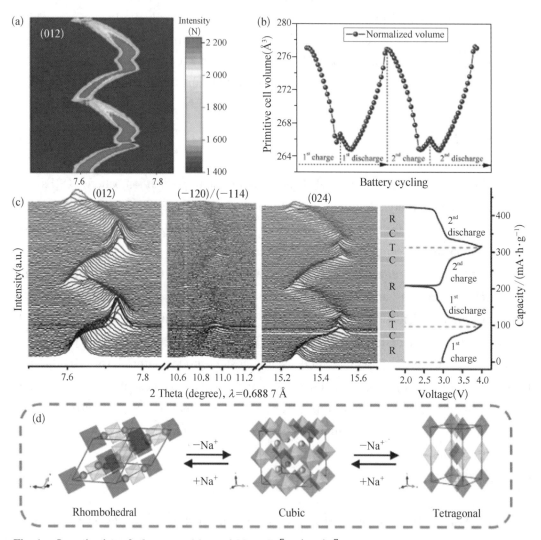

Fig. 1 Investigation of phase transitions of $Na_{1.73}Fe[Fe(CN)_6] \cdot 3.8H_2O$ sample with rhombohedral structure during cycling. (a) 2D contour plot of (012) reflection, (b) normalized volume during charge-discharge process obtained from synchrotron in-situ PXRD patterns of rhombohedral PB-S3, (c) (012), (−120)/(−114), and (024) reflections of synchrotron PXRD patterns, (d) schematic phase evolutions during cycling of rhombohedral PB-S3

information on Fe in the as-prepared samples. The concentrations of Na and Fe in the samples were measured by inductively coupled plasma (ICP) analysis (OPTIMA 8000DV Optical Emission Spectrometers).

【Results】

We have successfully synthesized sodium-rich $Na_{2-x}FeFe(CN)_6$ with a highly reversible rhombohedral structure via a simple scalable co-precipitation method, the nucleation, and growth process from nanoparticle to highly crystalline $Na_{2-x}FeFe(CN)_6$ microcube during precipitation is discovered. Synchrotron in-situ PXRD shows that the rhombohedral structure of $Na_{1.73}Fe[Fe(CN)_6] \cdot 3.8H_2O$ is highly reversible during the charge-discharge process with three-phase transitions between rhombohedral, cubic, and tetragonal. Benefiting from its stable structure, $Na_{1.73}Fe[Fe(CN)_6] \cdot 3.8H_2O$ sample shows excellent electrochemical performance with high initial Coulombic efficiency 97.4%, 70 mAh · g^{-1} discharge capacity was retained at current density 2 000 mA · g^{-1}, capacity retention was maintained at 71% after

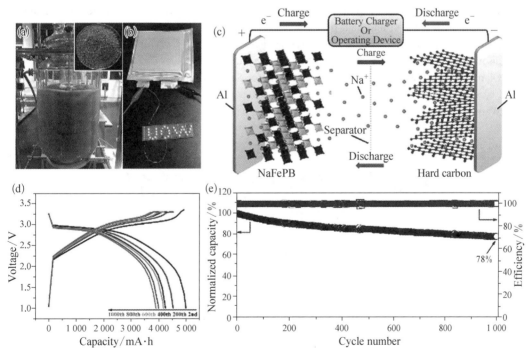

Fig. 2 (a) Digital image of synthesis of Prussian white $Na_{2-x}FeFe(CN)_6$ in 100 L reactor and powder of final product, (b) digital image of pouch full cell connected with red LED lights, (c) working mechanism of pouch full cell, (d) charge-discharge curves of pouch full cell, (e) cycling performance of pouch full cell

500 cycles. To demonstrate the practical application of rhombohedral $Na_{2-x}FeFe(CN)_6$, a scale-up co-precipitation experiment was conducted by using a 100 L reactor and a pouch full cell was fabricated as well, which shows a stable cycling performance over a thousand times. Our work might guide rational synthesis for other PBAs from scalable co-precipitation method, which would pave the way for mass-producing PBAs and designing high-performance SIBs in the future.

References

[1] Hwang J Y, Myung S T, Sun Y K. Sodium-ion batteries: present and future[J]. Chemical Society Reviews, 2017, 46: 3529-3614.
[2] Delmas C. Sodium and Sodium-Ion Batteries: 50 Years of Research[J]. Advanced Energy Materials, 2018, 8(17): 1703137.
[3] Kim H, Ding Z, et al. Recent Progress in Electrode Materials for Sodium-Ion Batteries[J]. Advanced Energy Materials, 2016, 6: 1600943.
[4] Yan Z, et al. A Hydrostable Cathode Material Based on the Layered P2@P3 Composite that Shows Redox Behavior for Copper in High-Rate and Long-Cycling Sodium-Ion Batteries[J]. Angewandte Chemie International Edition, 2019, 58(5): 1412-1416.
[5] Wang P F, You Y, Yin Y X, Guo Y G, et al. Layered Oxide Cathodes for Sodium-Ion Batteries: Phase Transition, Air Stability, and Performance[J]. Advanced Energy Materiacs, 2018, 8: 1701912.
[6] Chen M, Hua W, Xiao J, et al. NASICON-type air-stable and all-climate cathode for sodium-ion batteries with low cost and high-power density[J]. Nature Communications, 2019, 10(1): 1480.
[7] Yan G, et al. Higher energy and safer sodium ion batteries via an electrochemically made disordered $Na_3V_2(PO_4)_2F_3$ material. Nature Communications, 2019, 10: 585.
[8] Qian J, et al. Prussian Blue Cathode Materials for Sodium-Ion Batteries and Other Ion Batteries[J]. Advanced Energy Materiacs, 2018, 8(17): 1702619.
[9] Wang B, Han Y, Wang X, et al. Prussian Blue Analogs for Rechargeable Batteries[J]. iScience, 2018, 3: 110-133.

纤维电子器件的连续化制备与实用性评价研究

侯成义*

东华大学 纤维材料改性国家重点实验室
*通讯作者：hcy@dhu.edu.cn

【引言】

随着可穿戴技术的发展，柔性电子材料与器件获得了人们的大量关注，目前可穿戴领域的研究相继展示了"佩戴"形式的柔性器件,但多数仍缺乏柔韧性、不可拉伸、难以编织,其主要作为服装的附加品,缺乏穿着舒适性。

相比之下,纤维作为常用的服装材料是现成的电学功能物理载体,是更为理想的可穿戴功能集成平台。鉴于此,我们认为纤维将成为新一代的柔性器件形态。我们利用熔融纺丝和静电纺丝等连续化工艺,制得多组分多界面的电学纤维与织物,通过防水透气功能层的设计解决了器件性能易受环境湿度影响的问题,并研究了纤维电子器件的穿着舒适性。并通过演示实验验证了此类新型服装材料在随身能源与体征传感等领域的应用价值。

【结果】

采用熔融纺丝工艺,连续化制得千米级导电纱线@橡胶皮芯结构摩擦电纤维器件,通过橡胶防水功能层的设计,解决了电子器件怕水的问题。这种连续化制备、可编织的两栖电学纤维器件具有约 $1 \ W \cdot cm^{-2}$ 的峰值功率密度。

采用静电纺丝技术制备了铁电聚合物[P(VDF-TrFE)]和聚酰胺 6(PA6)两种纳米纤维作为功能材料,通过摩擦表面极化和铁电极化的相互作用,实现了摩擦/铁电协同电学增强。在低频外力作用下,这种电子织物材料可产生 $5.2 \ W \cdot m^{-2}$ 的峰值功率密度。

利用亲水聚丙烯腈(PAN)和聚酰胺 6 微/纳米纤维和疏水棉织物构筑了额外的吸湿排汗层,不仅能够有效地吸收体表汗液,其纳米纤维网络还可以明显加速汗液的扩散与蒸发,创造干燥的体感。全纤维的设计理念,保证了织物材料优良的透气和透湿性能,其较低的干态热阻和蒸发阻有利于维持舒适的体表微环境。

参考文献:

[1] Yang W, Gong W, Hou C, et al. All-fiber Tribo-ferroelectric Synergistic Electronics with High Thermal-moisture Stability and Comfortability[J]. Communications, 2019, 10(1): 5541.
[2] Gong W, Hou C Y, Zhou J, et al. Continuous and Scalable Manufacture of Amphibious Energy Yarns and Textiles[J]. Communications, 2019, 10(1): 868.
[3] Shi Q, Sun J, Hou C, et al. Advanced Functional Fiber and Smart Textile[J]. Advanced Fiber Materials, 2019, 1(18): 3-31.

Field-effect Control of Emergent Properties in Low Dimensional Quantum Materials

Shi Wu

Lawrence Berkeley National Laboratory and
University of California at Berkeley

【Introduction】

Low dimensional materials, such as nanotubes, graphene, and transition metal dichalcogenides (TMDs), have been a focus of interest in both fundamental research and device applications. The key to unveiling novel physics and realizing innovative device functionality in reduced dimensionality is externally controlled modulation of charge density by doping. However, conventional doping by field-effect from applied gate voltage through a solid dielectric layer has two main limitations. First, the maximum carrier concentration is restricted by dielectric breakdown. Second, the local control of carrier doping requires the fabrication of sophisticated nanostructures that introduce impurities and lack flexibility.

【Experimental】

Graphene and TMD thin film field-effect transistors (FETs) are fabricated using the dry pick-up transfer technique and standard nanofabrication processes. Alternative doping techniques are used to introduce ultrahigh carrier density or high-resolution local doping in the 2D FETs. Low-temperature transport measurements are conducted to characterize the emergent properties of all the doped devices.

【Results】

By employing and developing the EDLT technique, ultrahigh carrier density more than $10^{14} cm^{-2}$ can be accumulated in atomically thin 2D materials. As a result, ambipolar transistor operations and gate-induced superconductivity are realized in a series of TMD thin flakes including MoS_2, $MoSe_2$, $MoTe_2$, and WS_2. On the other hand, an e-beam doping technique is newly developed, which allows the reversible writing of complex and nonvolatile doping patterns in van der Waals heterostructures with high spatial resolution, high carrier density, and high mobility, even at room temperature. It thus offers an effective route to create nanoscale circuitry with designable electronic functions.

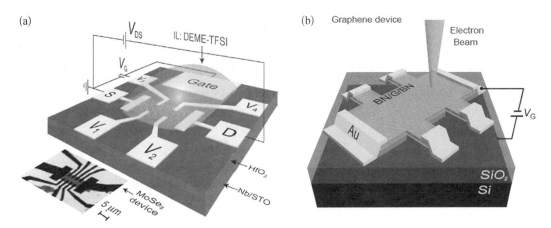

Fig. 1 Experimental scheme for (a) a MoSe$_2$ EDLT device and (b) e-beam induced doping in a BN-encapsulated graphene device

Stretchable Conductive Nonwoven Fabrics with Self-cleaning Capability for Tunable Wearable Strain Sensor

Li Qianming, Liu Hu*, Liu Chuntai, Shen Changyu

Key Laboratory of Materials Processing and Mold, Ministry of Education; National Engineering Research Center for Advanced Polymer Processing Technology, Zhengzhou University

* Corresponding author: 18437961095@163.com

【Introduction】

With the fast growth of wearable intelligent devices, tunable strain sensors with broad strain sensing range and high sensitivity are in urgent demand. Furthermore, the merits of the excellent waterproof property, self-cleaning, and anti-corrosion are also imperative for the practical applications of them.

【Experimental】

Herein, a tunable wearable micro-cracked non-woven fabrics (NWF) strain sensor with a broad strain sensing range and high sensitivity was successfully developed based on the electrically conductive cellulose nanocrystal (CNC)/graphene (G) coating with controllable micro-crack density, which was achieved through changing the G loading of the coating layer, and its superhydrophobicity was achieved by simply dip-coating in the hydrophobic fumed silica (Hf-SiO$_2$)/ethanol dispersion.

【Results】

As a result, a broad working range of 98% and GF value up to 2.36×10^4 are simultaneously achieved for the prepared strain sensor. As far as we know, they are the highest values ever reported. Besides, it possesses an ultralow detection limit as low as 0.1%, the short response time (33 ms), and good sensing stability over 1 000 cycles. What's more, the NWF strain sensor presents excellent waterproofness (WCA=154°),

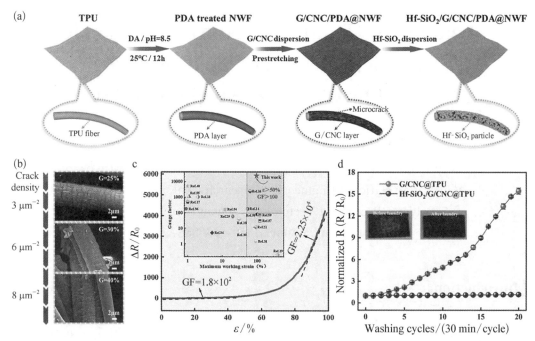

Fig. 1 (a) Schematic preparation of Hf-SiO$_2$/G/CNC/PDA@NWF composites. (b) Surface SEM images of the G/CNC/PDA@NWF composites with different G loading. (c) Comparison of the maximum working range and GF of the sensor in previous publications. (d) The resistance change of the strain sensor with and without Hf-SiO$_2$ depending on repeated washing cycles

anti-corrosion ability, outstanding self-cleaning, and stability. Due to the excellent response performances, the NWF strain sensor could be applied for the full-range human motion detection, including the large-scale (such as the bending of fingers, wrists, and arms) and small-scale (such as pulse, ultrasonic wave, and voice recognition) strain change, especially for the wet or rainy conditions.

This work provides a new perspective for fabricating a high-performance strain sensor with the tunable micro-crack structures and the self-cleaning capability. These results provide important progress towards the understanding of the role of microstructure in the realization of high sensitivity, broad response range, and superhydrophobic strain sensors.

Novel Solid-state Elastocaloric Materials for Eco-refrigeration

Xiao Fei, Jin Xuejun*

School of Materials Science and Engineering
Shanghai Jiao Tong University, Shanghai 200240
* Corresponding author: jin@sjtu.edu.cn

【Introduction】

Relatively low energy efficiency and emission of refrigerant which is harmful to the environment are the two issues inevitable for the most commonly used vapor-compression refrigeration technology. Elastocaloric effect (eCE) is a possible alternative technology because some alloys show large isothermal entropy change (ΔS_{iso}) and adiabatic temperature change (ΔT_{adi}) through stress application or removal. Ti-Ni alloys are the most promising eCE materials due to their large cooling ability and good workability. However, the fatigue behavior of Ti-Ni alloys must be improved. Here, we report that stable and giant eCE is obtained in nanocrystalline Ti-44Ni-5Cu-1Al (at%) alloy, which shows successive B2-B19-B19' martensitic transformation.

【Experimental】

An ingot of Ti-44Ni-5Cu-1Al (at%) alloy was prepared by vacuum induction melting in a graphite crucible and cast into an iron mold. A slab was cut from the ingot and was hot rolled at 1 123 K to 50% thickness reduction (the final thickness was ∼2 mm), and further cold rolled to 50% thickness reduction (the final thickness was ∼1 mm). Specimens used in the present study were cut from the cold-rolled sheet and annealed at 673 K for 5 min in evacuated quartz tubes followed by quenching into ice water. The mechanical tests was conducted using an INSTRON－5969 machine. The temperature change were monitored by FLIR SC7700M infrared camera and J-type thermocouple. The microstructural evolution was detected by a Rigaku RAPID II transmission type X-ray diffraction.

【Results】

The stress-strain curves and cooling ability of the Ti-44Ni-5Cu-1Al specimen hardly changed through 5 000 mechanical cycles (a), (b). The maximum adiabatic temperature decrease (ΔT_{adi}) by stress removal from 600 MPa was ∼25 K with a small temperature distribution of ∼0.5 K (c). The value of ΔT_{adi} is consistent with that calculated from strain-temperature curves using Maxwell relation (d). The effective working temperature window was ∼55 K, resulting in a high refrigeration capacity of

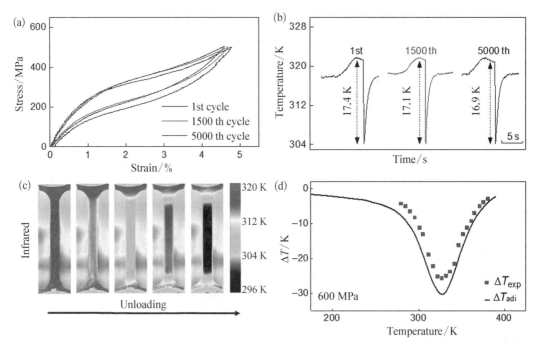

Fig. 1 (a) The stress-strain curves and (b) cooling ability for Ti-44Ni-5Cu-1Al; (c) the relationship between the adiabatic temperature and stress; (d) the value of the maximum adiabatic temperature from calculation and experiment

RC=4.2 kJ/kg. The material coefficient of performance reached COP $\sim$ 9.6 when an Otto cycle is considered. These properties are related to the diffuse nature of successive B2 – B19 – B19' martensitic transformation of this alloy.

Design and Development of Advanced Transparent Insulation Materials for Energy Efficient Windows

Ming Yang, Liu Xiao, Liu Xin, Sun Yanyi, Robin Wilson, Wu Yupeng[*]

Faculty of Engineering, The University of Nottingham,
University Park, Nottingham, UK, NG7 2RD
[*] Corresponding authors: Yupeng.Wu@nottingham.ac.uk;
Jackwuyp@googlemail.com

【Introduction】

In the EU, buildings are responsible for 40% of energy end-use. Windows, which provide vision, passive solar gain, and daylighting, are important parts of the building envelope. Meanwhile, they have significant effects on heating, cooling, and lighting energy consumption for buildings due to its relatively higher U-value compared to other components of building envelope and the transparency characteristic. Window integrated Transparent Insulation Material (TIM) is a potential solution to improve building energy efficiency through an increased window thermal resistance; and meanwhile, achieve a more comfortable luminous environment through scattered sunlight penetrations. However, daylight varies in intensity, colour, and direction over time, effecting building occupants' health and wellbeing. Intelligent solutions, which can regulate the changing environment, is highly anticipated. Combing TIM with thermotropic materials, which changes from transparent to translucent states when the temperature is above its transition temperature, provides the potential to dynamically respond to ambient conditions. This project aims to design and develop an advanced window integrated with Parallel Slat Transparent Insulation Material (PS-TIM) where the slats are made of thermotropic materials, providing a dynamic regulation for comfort luminous environment and energy-efficient building. Be more specific, the PS-TIM can reduce the conductive and radiative heat transfer through the window, therefore, increasing its thermal resistance. This can dynamically control the daylight rays through the window to achieve a uniform light distribution within the indoor space. In this project, a few thermotropic materials have been selected, and their optical (transmittance and reflectance in the entire wavelength) and thermal properties have been comprehensively studied. Suitable materials can be further selected for the proposed PS-TIM development.

【Experimental research methods】

PNIPAm and hydroxypropyl cellulose (HPC) have been selected for the proposed PS-TIM, PNIPAm was synthesized in our lab, and HPC was purchased from Sigma. The thermal and optical properties (transmittance, reflectance, transition temperature range, etc.) of the selected hydro polymers with different molecular weight and concentration were measured by using a spectrometer.

【Results】

From the experimental tests, it was found that the HPC hydro-membrane has a sharp hysteresis compared to that of the PNIPAm hydro-membrane during the phase change. In general, all the tested HPC samples with a molecular weight of 80 000, 370 000, 1 000 000 at a concentration of 1 wt %, 3 wt %, and 5 wt %, have transition temperatures in the range from approximately 39 ℃ to 50 ℃. This is only within 1℃ for PNIPAm, which has a transition temperature at approximately 32℃. It was also found that under the same Mw, the HPC membrane with higher concentration is more sensitive to temperature than the low concentration one. This is similar to HPC under different Mw. Besides, at the same Mw level, the transmittance is similar before switching under various wt.%. a Higher wt.% sample offers lower transmittance (This gives a maximum difference $\Delta\tau_{max}$ over 15% between the 1 wt% and 5 wt% HPC) and higher reflectance (maximum difference $\Delta\rho_{max}$ can be over 15%) after switching. Besides, under the same wt.%, with an increase Mw from 80 k to 1 000 k, the reflectance increase after switching, transmittance decrease with increase Mw. This only happens for the wt.% below 3 wt.%. The maximum transmittance difference $\Delta\tau_{max}$ is over 20% for HPC and $\Delta\tau_{max} > 15\%$ for PNIPAm when transition completed. However, the maximum reflectance differences $\Delta\rho_{max}$ are only over 10% for both HPC and PNIPAm. Meanwhile, the $\Delta\rho_{max}$ decreases with the increase in concentration. It can be concluded that both two thermotropic materials with higher wt.% and Mw have better control of glare caused by strong sunlight, and potentially provide a comfortable indoor luminous environment. PNIPAm has a better response to the stimulus of temperature compared with that of HPC and is more suitable for the proposed PS-TIM system.

 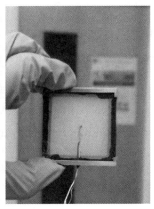

(a) Membrane state above T_s (b) Membrane state below T_s

Fig. 1 Photo of the developed thermotropic material membrane

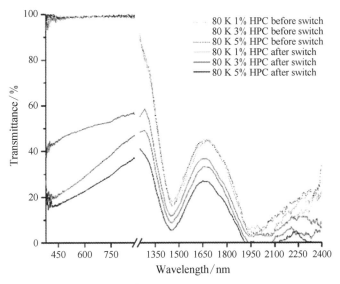

Fig. 2 Different wt.% 80 000 Mw HPC solution membrane spectra showing transmittance change

Smart Nanothermochromic Window Advances and Implications for Use in Different Climates

Marina Aburas[1,2], Zhao Jiangbo[1], Terence Willimason[1], Veronica Soebarto[1], Heike Ebendorff-Heidepriem[1], Wu Yupeng[2]*

1 University of Adelaide (Australia) 2 University of Nottingham (United Kingdom)
* Corresponding author: yupeng.wu@nottingham.ac.uk

【Introduction】

Thermochromic windows come with a promise of energy-saving capabilities. Concerning the energy performance of these windows, previous research has found that their energy performance varies significantly (up to 73.4%) between different cities using the same film type but less (up to 21.6%) between different films tested in the same city. This presentation reports on research that investigated the influence of nanothermochromic coating thickness on transmittance, solar modulation, and, consequently, energy-saving performance of smart windows in different climates: Desert, Mediterranean, Temperate, and Subarctic.

【Experimental】

VO_2(M) nanoparticles have exhibited high luminous transmittance and infrared (IR) modulation capability. The crystal type, composition, and size of the nanoparticles, around 30 nm produced via a hydrothermal method, were determined using x-ray powder diffraction (XRD) and scanning electron microscopy (SEM), respectively. The VO_2 nanoparticle powder was suspended so that a nanoparticulate film could be made via spin-coating of the suspension. The monoclinic powder was incorporated into an HCl/ethanol suspension, which was then used for preparing a thin film of VO_2 onto a soda-lime-silicate substrate in ambient conditions. Characterization techniques of the thin-films included XRD, SEM, optical profiling, IR imagery, and spectroscopy. The thin films, with varying layer thicknesses (Fig. 1), displayed thermochromic switching function through a distinct semiconductor-to-metal transition. The developed thermochromic windows were investigated for application in an office building model using a validated simulation engine EnergyPlus.

【Results】

The in-house produced nanothermochromic thin films resulted in a variation in luminous transmittance (T_{lum}) between 30% – 75% as well as infrared

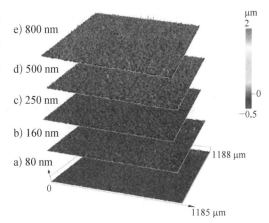

Fig. 1 Surface roughness profiles with varying coating thicknesses thermochromic thin-film layers

modulating capability (ΔT_{IR}) between 2% – 6% (Fig. 2), which was achieved by adjusting the thickness of the spin-coated film composites. The advantages of this facile method are simplicity of the process, high purity of starting materials, compatibility with a wide range of substrates, and reproducibility in switching properties, making the method practical for smart window applications. The simulation results showed that the thicker film with $T_{lum}=30\%$ produced the highest energy saving characteristic compared with double glazing in all climates (Fig. 3). The highest relative energy saving over a tinted window in the Dessert climates also occurs for this film, whereas in Subarctic climates this occurs with thinner films of $T_{lum}=75\%$. These smart windows also reduced peak load and visual discomfort in all climates. Overall results indicated that with comprehensive evaluation for optimal selection,

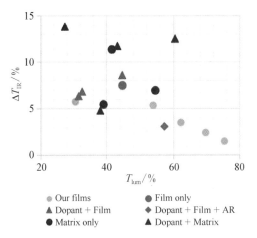

Fig. 2 Tlum vs ΔTsol and for various thermochromic window types

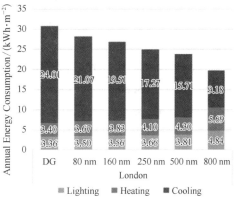

Fig. 3 Annual Energy Consumption of thermochromic windows modelled for a Temperate climatic condition

thermochromic glazing can reduce energy use to varying extents depending on film coating thickness and climate.

Ultralight Programmable Bioinspired Aerogels with Integrated Multifunctionalities via Co-assembly

Cai Chenyang, Fu Yu*

Nanjing Forestry University
* Corresponding author: fuyu@njfu.edu.cn

【Introduction】

Creating a configurable and controllable surface for structure-integrated multi-functionality of ultralight aerogels is of significance but remains a huge challenge due to the critical limitations of mechanical vulnerability and structural processability. Herein, inspired by Salvinia minima, the facile and one-step co-assembly approach is developed to allow the structured aerogels to spontaneously replicate Salvinia-like textures for function-adaptable surfaces morphologically. By introducing the binding groups for hydrophobicity tailoring, functionalized nanocellulose (f-NC) is prepared via mechanochemistry as a structural, functional, and topographical modifier for a multitasking role. The self-generated bioinspired surface with f-NC greatly maintains structural unity and mechanical robustness, which enable self-adaptability and self-supporting of surface configurations. With fine-tuning of nucleation-driving, the binary microstructures can be controllably diversified for structure-adaptable multi-functionalities. The resulting ultralight Salvinia minima-inspired aerogels (e.g., 0.054 g·cm^{-3}) presented outstanding temperature-endured elasticity (e.g., 90.7% high-temperature compress-recovery after multiple cycles), durable superhydrophobicity and anti-icing properties, oil absorbency efficiency (e.g., 60.2 g·g^{-1}), thermal insulating (e.g., 0.075 W·mk^{-1}), which are superior to these reported on the overall performance.

【Experimental】

With tuning hydrophobicity of nanocellulose modifier, the on-demand microphase invasion was controllably initiated because of different wettability between the waterborne polyurethane (WPU) matrix and partially superhydrophobic nanocellulose. Subsequently, the occurring nonsolvent induced microphase separation allowed for the transient localization of colloidal blocks to form the heterogeneous growth points of ice nucleation. With the help of a cryogenically cooled driving force, highly localized ice fronts were regularly arranged and directionally evolved at the top of the superhydrophobic modifier. After sublimation, the colloidal building blocks were synergistically coalesced and directionally solidified for the topographically patterned monolithic structures with intrinsic topological geometries. By controlling velocities of the heterogeneous ice growth, the predesigned topography was in-situ superimposed on intrinsic topological hierarchy for bioinspired binary microstructures with desirable dual wettability.

【Results】

It could be found that WPU and a-WPU showed a strong amphiphilic characteristic because both the water and the gasoline were absorbed into their interior structures rapidly. In contrast, the droplets of water remained onto the surface of the f-WPU, but the droplets of gasoline were rapidly immersed into the interior structures, demonstrating the engineered dual wettability of superhydrophobicity and superoleophilicity. The obtained aerogels could withstand a compressive strain of 40% and almost recovered its original shape after the release of the stress after 3 cycles at room temperature. Compared with other WPU based aerogels via freeze casting, the developed f-WPU exhibited comparably superior compress strength and elasticity (after 10 cycles). The high porosity and assembly-directed hierarchy of as-prepared aerogels could lead to extremely low gas and solid thermal transportation. Comparing with the commercially available thermal insulating materials such as PU aerogels, PI foams, Wood, PVC tubes, etc. the f-WPU exhibited much lower thermal conductivity and extremely low density of $0.054 \text{ g} \cdot \text{cm}^{-3}$, the lowest among most thermally insulating materials reported.

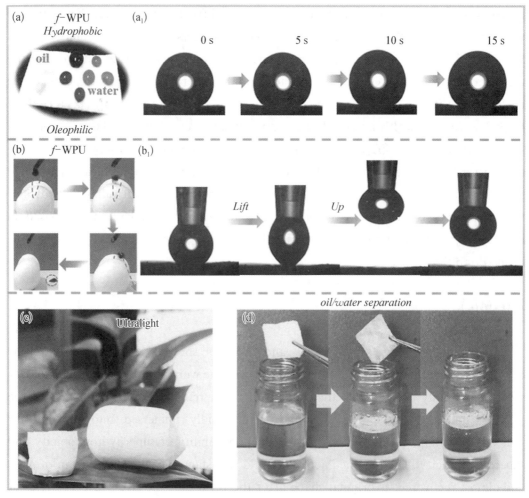

Fig. 1　Wetting property of f-WPU aerogel(Sample)

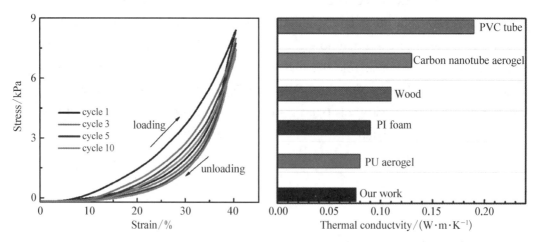

Fig. 2　Mechanical property and thermal insulating of f-WPU aerogel (Sample)

第五部分　软物质材料
Section 5　Soft Mater Materials

Unlocking Polymer Degradability Through Mechanophore Activation

Zhou Junfeng, Hsu Tze-gang, Su Hsin-wei, Wang Junpeng*

Department of Polymer Science, The University of Akron

*Corresponding author: jwang6@ukaron.edu

【Introduction】

Synthetic polymer materials usually exhibit high chemical stability and heat resistance. These advantages make them difficult for degradation in the natural environment also well, resulting in environmental pollution. The developing degradable polymer is one of the effective ways to reduce such plastic pollution. In this case, polymers can be endowed with degradability by incorporation of functional groups (such as esters, acetals, silyl ethers) into the polymer backbones. However,

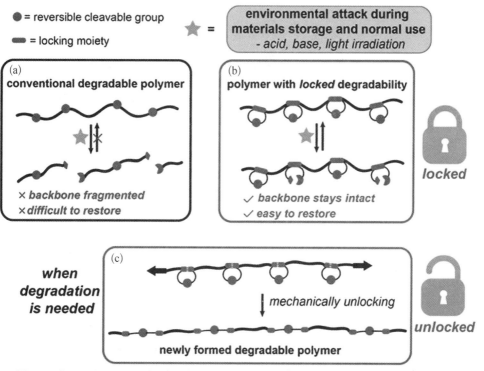

Fig. 1 Comparison diagram of degradable polymers and polymers with locked degradability

these groups also reduce the durability of the polymers because they can be triggered by environmental cues, which is detrimental to the mechanical properties of polymers. Although modifying the substituents near the cleavable groups, tuning the tacticity and crystallinity, and adding stabilizers can enhance the stability of degradable polymers caused by cleavable functional groups, but these methods are usually limited by the structure of polymers. Therefore, a more general strategy is needed to improve the stability of the polymer while still maintaining its degradability.

【Experimental】

The lactone monomer (M1) was facile prepared through the photochemical[2+2] cycloaddition of maleic anhydride with cyclooctadiene and then reducing by $LiAlH_4$. The polylactone (P1) was obtained by ROMP (Ring Opening Metathesis Polymerization) in which the cleavable lactone motif is fused with cyclobutane that enables degradability is locked by the cyclobutane mechanophore. The stability and degradability of the P1 before and after sonicated were characterized by ^{1}H NMR, GPC, and mass spectrometry. The P1 can be further hydrogenated to a saturated polymer (HP1). The thermal properties of P1 and HP1 were characterized by DSC and TGA.

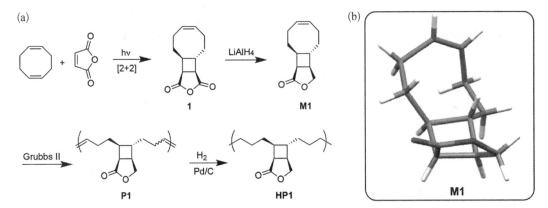

Fig. 2 (a) Synthesis of the polylactone; (b) Single-crystal structure of monomer

【Results】

Highly stable cyclobutane keeps the polymer backbone intact when hydrolysis of lactone occurs. Confirmed by the results of ^{1}H NMR and GPC, the molecule weight remained unchanged when 90% of lactones were hydrolyzed under basic conditions.

Moreover, the hydrolyzed polymer can be re-esterified to restore the pristine lactone structure. When degradation is required, the cyclobutane mechanophores can be activated and undergo electrocyclic ring-opening reactions by sonication to induce the resulting ester groups into the polymer backbone, unlocking the degradability. This new class of polymer addresses the trade-off issue between stability and degradability, allowing that the degradability is lacked during storage and use, and unlocked on demand.

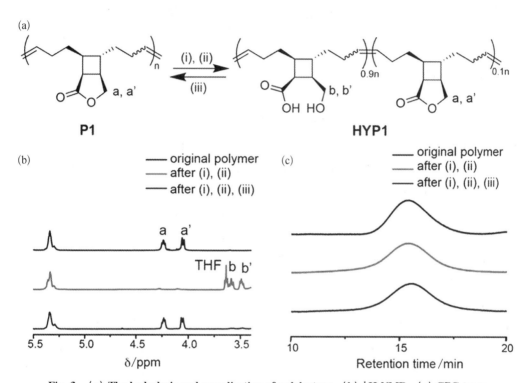

Fig. 3 (a) The hydrolysis and recyclization of polylactone; (b) ^{1}H NMR; (c) GPC trace

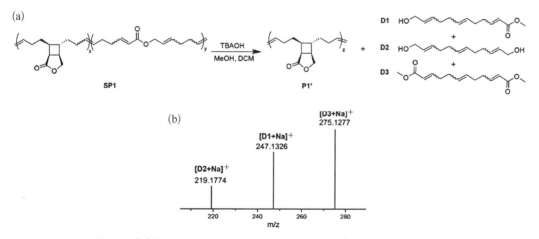

Fig. 4 (a) Methanolysis of the sonicated polymer (SP1) with TBAOH;
(b) The mass spectrum of the degradation products

Design of Polymers for Biomedical Applications

Ren Kaixuan*

University of Southern California

* Corresponding author: kaixuanren2015@gmail.com

【Introduction】

Articular cartilage is an organized tissue that allows for smooth motion in diarthrodial joints. This tissue possesses distinct load-bearing and low-friction capabilities, which benefit from its extracellular matrix (ECM) composed of water, collagen, and glycosaminoglycans (GAGs). Damage to the articular cartilage can be caused by trauma, disease, or sports-related injury, leading to disability of joint function. Without access to the blood supply, abundant nutrients, and progenitor cells, the damaged articular cartilage cannot repair and regenerate itself. Current clinical therapies for articular cartilage defects include autologous chondrocyte transplantation (ACI), osteochondral auto-and allografts, and microfracture. Although these techniques have shown some efficacy in remodeling cartilage defects, the therapies have limitations such as the lack of donors, poor integration to the surrounding cartilage tissue, and formation of fibrous tissue instead of hyaline cartilage.

Recently, cartilage tissue engineering provides a promising approach for the repair and regeneration of cartilage defects. Herein, I developed an injectable hydrogel consisting of poly (l-glutamic acid)-*graft*-tyramine (PLG-*g*-TA) with a tunable microenvironment and used as an artificial ECM to investigate the influence of 3D matrix microenvironment on the behaviors of bone-marrow-derived mesenchymal stem cells (BMSCs) during three-dimensional (3D) culture. Moreover, I designed glycopolypeptide-based injectable hydrogels as biomimetic scaffolds to act as an analog of proteoglycans present in the ECM of native cartilage for cartilage tissue engineering.

【Experimental】

The physicochemical properties of the hydrogels, such as the mechanical properties, morphology, swelling, and degradation behaviors, were measured by rheometer, SEM, and degradation experiments. The cytocompatibility of the hydrogels was measured by MTT assay. The biodegradability and biocompatibility of the hydrogels *in vivo* were evaluated by subcutaneous injection of the hydrogels into rats. To assess the feasibility of the hydrogels as 3D scaffolds for cartilage tissue engineering, rabbit BMSCs or chondrocytes were incorporated in the hydrogels, and the cell viability and proliferation within the hydrogels were examined by live-dead assay and cell counting kit-8 method *in vitro*. Additionally, the morphology of cells and formation of the cartilaginous specific matrix within the hydrogels were investigated.

【Results】

It is intriguing to note that the BMSCs within the 2% hydrogel showed a well-spread morphology after 24 h and a higher proliferation rate during 7 days of culture, in contrast to a rounded morphology and lower proliferation rate of BMSCs in the 4% hydrogel. Furthermore, the hydrogels with different microenvironment also regulated the matrix biosynthesis and the gene expression of BMSCs. After incubation in the 2% hydrogel for 4 weeks, the BMSCs produced more type II collagen and expressed higher amounts of chondrogenic markers, compared to the cells in the 4% hydrogel. Therefore, the PLG-*g*-TA hydrogels with tunable microenvironment may serve as an efficient 3D platform for guiding the lineage specification of BMSCs. The chondrocytes cultured in the glycopolypeptide hydrogels showed a spherical phenotype

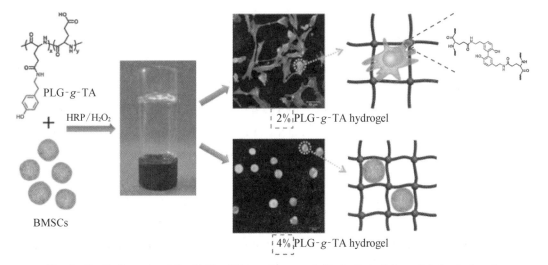

Fig. 1 Synthetic route of the PLG-*g*-TA copolymer and illustration of the cell-laden hydrogels

with high viability during the culture *in vitro* for up to 21 days and displayed a higher proliferation rate than in the control hydrogel counterpart. The chondrocytes encapsulated in the glycopolypeptide hydrogels incubated *in vitro* as well as in the subcutaneous model of nude mice maintained chondrocyte phenotype and formed the cartilaginous specific matrix. Therefore, the biomimetic, biodegradable glycopolypeptide-based hydrogels may serve as promising 3D scaffolds for cartilage tissue engineering.

Polypeptide Superhelices: Chirality and Morphology

Cai Chunhua*, Lin Jiaping

School of Materials Science and Engineering, East China University of Science and Technology, Shanghai 200237

* Corresponding author: caichunhua@ecust.edu.cn

【Introduction】

Block copolymers are capable of self-assembling into diverse nanostructures. Chiral assemblies, for example, superhelix have been attracted increasing attention in recent years. Controlling of the chirality as well as the morphology of the superhelix is the core science of this research field.

【Experimental】

In this work, we reported that poly(γ-benzyl L-glutamate)-*block*-poly(ethylene glycol) (PBLG-*b*-PEG) polypeptide block copolymer and PBLG homopolymers can cooperatively self-assemble into superhelices (Fig.1a). The superhelices were prepared by first dissolving the PBLG-*b*-PEG and PBLG homopolymers in a mixture solvent of tetrahydrofuran/N, N'-dimethylformamide (THF/DMF), then water was dropped into the solution to induce self-assembly of the polymer mixtures. Finally, after dialyzing against water, an aqueous solution of superhelices was obtained. By a combination of experiment and simulation, a core-shell structure of these superhelices was revealed, in which PBLG homopolymer bundles serve as the core of the superhelices, while PBLG-*b*-PEG block copolymers helically warp into screws of the superhelices. The influencing factors controlling the chirality and morphology of the polypeptide superhelices are studied.

【Results】

From two fundamental aspects, *i.e.* the chirality and morphology of the superhelices, we carried out systematical researches. (1) The chirality (right-handed or left-handed) of the superhelices can be controlled by the nature of the initial solvent and the preparation temperature (Fig.1b). When the volume fraction of DMF in the initial THF/DMF mixture solvent increases or the preparation temperature is higher, left-handed superhelices are preferred. In such chirality transitions, the chirality of the PBLG backbone does not change, the chirality transition is induced by the chiral arrangement of the pending phenol groups. Also, the chirality of the superhelices can be varied by the chirality of the polypeptide blocks. (2) Dependence of the morphology of the superhelices on the molecular weight of the PBLG homopolymers was observed (Fig.1c). When the molecular weight of the PBLG homopolymer is lower, pure rod-like superhelices are formed; when the molecular weight of the PBLG homopolymer is higher, pure toroidal superhelices are formed.

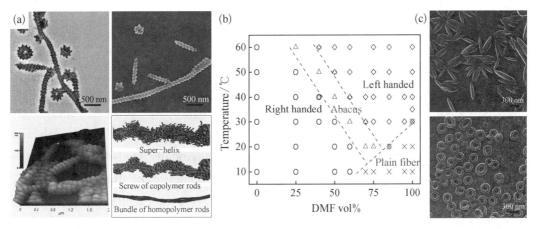

Fig. 1 (a) Morphology and structure of the superhelices self-assembled from PBLG-*b*-PEG/PBLG mixtures; (b) Chirality phase diagram of the superhelices as a function of the solvent nature and temperature; (c) Identical rod-like and toroidal superhelices

Renewable and Degradable Polymeric Materials Based on Coordination Ring-opening Polymerization

Zhu Yunqing*

Department of Polymers, School of Materials Science and Engineering, Tongji University, 4800 Caoan Road, Shanghai, PRC, 201804
* Corresponding author: 1019zhuyq@tongji.edu.cn

In recent years, many types of renewable polymers have been prepared using various types of renewable resources, but few have been able to combine their material properties (including thermal, mechanical, degradability, etc.), so they are often unable to match existing petroleum-based polymers. To address this issue, the following contents will be included in this presentation:

(1) Using new bimetallic (zinc) organic catalysts, a chemoselective coordination ring-opening polymerization method was proposed. This method can construct renewable multiblock copolyesters from a mixed monomer feedstock of cyclic anhydrides/epoxides/cyclic esters, which greatly simplify the synthesis route of multiblock polyesters. The thermodynamic mechanism of the chemoselective coordination ring-opening polymerization method was also investigated.

(2) The chemoselective coordination ring-opening polymerization mentioned above was then employed to prepare renewable and degradable functional polyesters with multi-block chain architecture. Then the impact of both the chain length and block ratio on the microphase separation was systematically studied. Through the regulation of the chain length, block ratio, and sequence of the multi blocks, functional polyesters, such as thermoplastic elastomers with excellent elasticity and shape memory polymers with a significant degree of recoverable deformation, were successfully prepared and patented.

(3) Via the coordination ring-opening polymerization method, an amphiphilic block copolyester was prepared with potential biomedical applications. The monomers used to synthesize this polyester were not only renewable but also derived from metabolic intermediates. After being degraded by enzymes in the cell, one of its degradation products, maleic acid, enters the Krebs cycle and indirectly regulates cell

behavior by accelerating cell metabolism and promoting proliferation. Therefore, when being self-assembled into vesicles, the copolyester exhibits excellent synergetic effect with the loaded drugs by regulating cell behavior to maximize the efficacy of these drugs.

Polymer Brush Based Inorganic-Organic Hybrid Materials

Yan Jiajun[1,2,*], Krzysztof Matyjaszewski[1]

1 Carnegie Mellon University
2 Present address: Lawrence Berkeley National Laboratory
* Corresponding author: yjjart@gmail.com

【Introduction】

Surface-modification of inorganic materials with polymer brushes merges the compatibility, versatility, and processability of polymeric materials and thermal, mechanical, optical, electric, and catalytic properties of inorganic materials. A range of methodologies, including silane, catechol, and phosphonate chemistries has been developed to functionalize inorganic surfaces for subsequent surface-initiated polymerization. However, each of these methods had a limitation in cost, accessibility of chemicals, and unfavorable side-reactions. We herein developed a novel method, inspired by fatty acids, universally applicable to a wide spectrum of metal and metal oxides, a major class of inorganic materials, surfaces.

【Experimental】

The tetherable initiator, 12-(2-bromoisobutyramido)dodecanoic acid (BiBADA), was prepared by reaction of ω-aminolauric acid and 2-bromoisobutyryl bromide in the presence of triethylamine. Inorganic particles were modified by the initiator in tetrahydrofuran under ultrasonication. Typical surface-initiated atom transfer radical polymerization (ATRP) was performed from initiator-modified particles using copper bromide as a catalyst, tris(2-aminoethyl)amine as a ligand, copper wire or tin(II) 2-ethyl hexanoate as a reducing agent, and anisole as a solvent.

【Results】

The tetherable initiator, BiBADA, comprised of a carboxylic acid head, an aliphatic backbone with 11 methylene units, and an ATRP initiating end-group, attaches to the metal oxide surfaces via coordinating bonds and solvophobicity, in a similar way to oleic acid, which is the most widely applied stabilizer for metal oxide nanoparticles. BiBADA was applied to metal or metal oxide surfaces in various approaches: (1) it was used to functionalize pristine metal oxide nanoparticles by ultrasonication in an organic solvent; (2) it was used as a stabilizer in metal oxide nanoparticle synthesis; (3) it was also directly cast on metal sheets from a solution. The resulting surface became compatible with organic solvents and ready for polymerization of various monomers (Fig. 1).

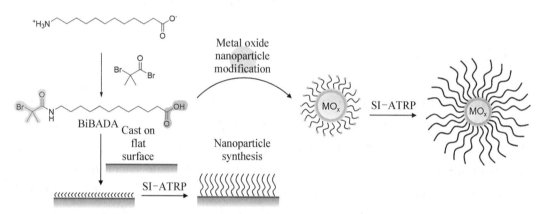

Fig. 1   Preparation of polymer brush-based inorganic-organic hybrid materials using fatty-acid inspired tetherable initiators

A broad range of metal oxide particles, including MgO, TiO_2, Co_3O_4, NiO, ZnO, Y_2O_3, ZrO_2, La_2O_3, CeO_2, WO_3, $\alpha\text{-}Al_2O_3$, In_2O_3, indium tin oxide, SnO_2, Sb_2O_3, $BaTiO_3$, etc., were successfully grafted with polymer brushes, among which many were never modified with polymer brushes before.

Many applications were developed for materials prepared using this novel technique (Fig. 2). For example, poly (methyl methacrylate) (PMMA) modified ZrO_2 nanoparticles were found to be stable over weeks in organic solvents. The solvent-cast hybrid film was twice as stiff as PMMA and one order of magnitude harder. Polymer brushes can be applied to macroscopic metal sheets with an intrinsic oxide layer, such as aluminum or steel, using a simple "paint-on" technique to drastically change the surface properties.

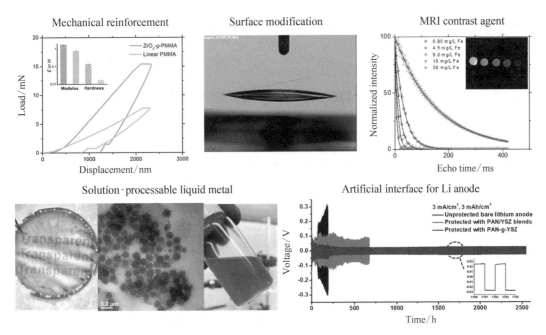

Fig. 2 Selected applications of polymer brush-based inorganic-organic hybrid materials using fatty-acid inspired tetherable initiators

Additionally, it finds applications in medications. Superparamagnetic iron oxide nanoparticles prepared with *in-situ* initiating groups were grafted with dimethyl sulfoxide-based polymer brushes. The water-dispersible hybrid particles were found to significantly boost the contrast of magnetic resonance imaging while bearing low cytotoxicity. Titanium-based dental implants were grafted by biocompatible and reactive monomers to allow conjugation of P15 peptides, which enhanced the osseointegration of the implants.

These hybrid materials were also applied in energy conversion and storage. Piezoelectric $BaTiO_3$ and TiO_2 particles grafted with PMMA were able to convert ultrasonic agitation into electrons and catalyze mechanochemical reactions. Similarly, matrix-free $BaTiO_3$-PMMA hybrid film showed a drastic improvement in permittivity while retaining the high breakdown strength as well as the mechanical integrity of PMMA. Eutectic gallium indium liquid metal nanodroplets covered by polymer brushes were readily processable from organic solvents or water into glassy or elastic materials. The liquid metal-polymer hybrid materials were made into thermoelectric wearable devices operable at sub-zero temperatures. Yttria-stabilized zirconia nanoparticles grafted with polyacrylonitrile was made into an artificial interface layer with high Li^+ conductivity, high transference number, and high modulus, which is promising to promote the performance and safety of Li metal batteries.

A large number of applications of the novel polymer brush-based inorganic-organic hybrid materials have been explored but there are many more yet to be discovered.

Tetrapod Polymersomes

Xiao Jiangang, Du Jianzhong*

Tongji University

*Corresponding author: jzdu@tongji.edu.cn

【Introduction】

The rich functionality and geometries found in the complex structures produced by nature are a continuous source of inspiration for many generations of scientists to design nature-inspired materials with beautifully-defined subtle nanostructures with applications in our society. Inorganic tetrapod nanocrystals have shown excellent performance and can be synthesized from a variety of inorganic materials through certain crystal growth techniques. "Colloidal molecules" with tetrapod-type structures can be prepared from controlled phase separation of immiscible macromolecules. However, most multipod nanomaterials are based on dense crystals or polymers without enough storage space for functional guest molecule loading, restricting further practical applications. At present, the design and preparation of biomedically-useful tetrapod polymersomes (also known as polymer vesicles) remain important challenges.

【Experimental】

DLS studies, DSC analysis, TEM images, SEM studies, and confocal laser scanning microscopy studies.

【Results】

We report the preparation and the formation mechanism, via fusion-induced particle assembly (FIPA), of an unusual tetrapod polymersome nanoparticle morphology (Fig.1) with four tetrahedrally distributed hollow pods that are, in nature, elongated polymersomes. The hydrophobic TBA and DEA moieties form the membrane, whereas the hydrophilic PEO chains form the coronas. These novel

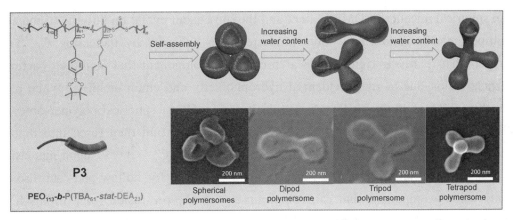

Fig. 1　Evolution of tetrapod polymersomes via fusion-induced particle assembly (FIPA)

nanoparticle species were obtained via the controlled fusion of four traditional spherical polymersomes as a result of simply changing the DMF/water ratio of the solvent. The original traditional polymersomes were self-assembled from PEO_{113}-b-$P(TBA_{61}$-$stat$-$DEA_{23})$ block copolymer (P3), which was synthesized by one-step reversible addition-fragmentation chain transfer (RAFT) polymerization.

To unravel the secret behind the tetrapod polymersomes, a series of block copolymers (P1 to P5, Fig. 2) with various comonomer types and degrees of polymerization were synthesized and self-assembled. PEO_{113}-b-$PTBA_{80}$ self-assembles

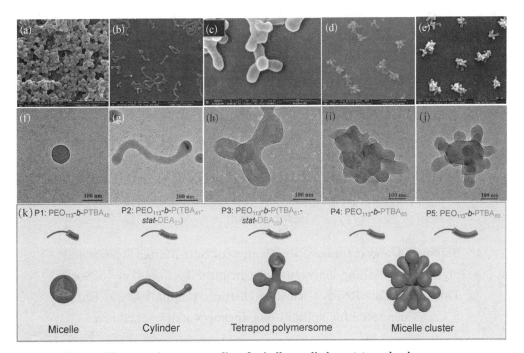

Fig. 2　Electron microscopy studies of micelles, cylinders, tetrapod polymersomes and micelle clusters self-assembled from copolymers P1 to P5

into spherical micelles in DMF/water, and the subsequent evolution into the tripod and multipod micelles and finally micelle clusters was achieved by increasing C_w. This suggests that relatively rigid TBA is a "pro-fusion" component that facilitates particle-particle fusion due to its providential hydrophobicity and chain mobility. When one-fourth of TBA of PEO_{113}-b-$PTBA_{80}$ is substituted by DEA, spherical polymersomes of PEO_{113}-b-$P(TBA_{61}$-$stat$-$DEA_{23})$ are born in DMF/water and then fused into dipod, tripod (C_w=95%), and finally tetrapod polymersomes (C_w=100%) upon increasing C_w, suggesting that flexible DEA is not only a promoter for hollow pods but also an "anti-fusion" component that can compromise with the pro-fusion force for its high chain mobility. The formation of either tetrapod polymersomes or micelle clusters is a matter of balance between pro-fusion and anti-fusion forces.

In conclusion, we propose a novel fusion-induced particle-assembly strategy for creating tetrapod polymersomes. The formation of either tetrapod polymersomes or micelle clusters is a matter of balance between pro-fusion and anti-fusion forces, which can be regulated by the comonomer types, copolymer composition, and solvent composition during the self-assembly/nanoparticle fusion process. In principle, FIPA can be extended to generate other hollow multipod nanostructures when the following key criteria are satisfied: (1) the original vesicle coronas should facilitate vesicle-vesicle adhesion (e.g., PEO coronas are suitable); (2) there is optimized membrane tension to trigger the fusion of vesicles (as achieved in this work by manipulating the ratio of rigid TBA to flexible DEA). Overall, our findings open up a new avenue for creating tetrapod polymersomes as well as other higher-order nanostructures with precisely defined spatial compartments.

Polymers/Carbon Nanomaterial Composites and Electrochemical Energy Storage

Geng Jianxin*

Beijing Advanced Innnovation Center for Soft Matter Science and Engineering, Beijing University of Chemical Technology, 15 North Third Ring East Road, Chaoyang District, Beijing 100029, China

*Corresponding author: jianxingeng@mail.buct.edu.cn

With the rapid developments of the markets of portable electronics and electric/

hybrid vehicles, efficient energy storage, and conversion materials with high power, high energy, and long life span are urgently needed. The rational design of polymer composites is an effective approach to develop high-performance energy materials. With the unique structures and outstanding physical and chemical properties, carbon nanomaterials including graphene and carbon nanotubes are the ideal candidates for energy storage and conversion materials. In our research, we focus on the fundamental issues such as the surface modification of carbon nanomaterials, the manipulation of multi-scaled structures of the composites as well as the relationship between the structures of the composites and the performance of the energy storage and conversion devices. Specifically, (1) a series of new approaches (including covalent, non-covalent, and surface-initiated polymerization strategies) have been developed for the preparation of graphene materials with versatile properties; (2) assembly structures of graphene sheets, as well the optimized morphologies of electroactive materials in graphene-based composites, have been prepared to enhance the specific capacitances of supercapacitors; (3) as for the use in electrodes of lithium batteries, a variety of new strategies that utilize covalent bonds, coordination bonds, and donor-acceptor interactions to enhance the interface properties in the carbon-based electrodes have been proposed to enhance the cycling stability of devices.

Energy Storage and Conversion Using Conjugated Microporous Polymers

Liao Yaozu*

Donghua University
* Corresponding author: yzliao@dhu.edu.cn

Efficient energy storage and conversion technologies represented by supercapacitors and hydrogen energy produced by water splitting are regarded as one of the key technologies for the development of clean energy. Despite the different technical principles, the development of photo-and electro-active materials with high performance and excellent stability is critical to their applications. Conjugated microporous polymers (CMPs) are a type of organic porous material linked by covalent bonds, featuring the characteristics of flexible structure design, free adjustment of pore and photoelectric activity, and rich modification of pore surface. The CMPs have been showing great promising in green energy development. We have developed a variety of coupling reactions to prepare such polymeric materials and their derived carbons, which showing unique advantages in electrochemical energy storage and conversions such as supercapacitors, hydrogen production, and fuel cells.

聚合物分子刷及其功能材料[①]

冯纯,李永军,陆国林,黄晓宇*

中国科学院上海有机化学研究所,中国科学院有机功能分子合成与
组装化学重点实验室　上海市零陵路 345 号,邮编 200032
* 通讯作者：xyhuang@sioc.ac.cn

一维、二维和三维聚合物分子刷是指聚合物链分别接枝到线性聚合物、平面基质和球形粒子表面上形成的独特聚合物体系。一维聚合物分子刷也称为接枝共聚物,它是指聚合物侧链密集接枝到线性聚合物链上所形成的共聚物。一维聚合物分子刷所具有的紧凑的结构可产生一些独特的性质,如蠕虫状构象、紧凑的分子尺寸和显著的链端效应。二维和三维聚合物分子刷是指聚合物链密集地链接在各种有机或无机基质表面上

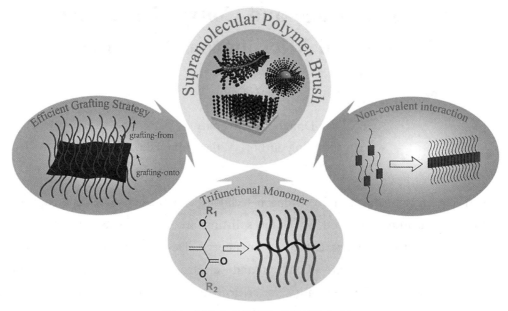

图 1　聚合物分子刷的高效可控合成

[①] 经费资助：国家杰出青年科学基金(No. 51825304)。

所形成的聚合物复合体系。该聚合物复合体系不仅保留了基质所具有的性质，聚合物链的引入还赋予了复合体系特殊的性质，如防腐蚀性、胶体稳定性、抗黏附性能、刺激响应性、润滑性和摩擦性等。但是聚合物分子刷的高效可控制备仍然是该领域的一个关键问题，某种程度上仍是一个挑战。另外，也需要发展性能优异的聚合物分子刷基有机高分子功能材料。

我们将有机化学和高分子化学相结合，开发了基于"三重功能性单体"为核心的高效合成一维聚合物分子刷的新策略。还发展了通过非共价途径-活性结晶驱动自组装来构造非共价键一维聚合物分子刷。在探究一维、二维、三维聚合物分子刷的可控制备的基础上，还探究了聚合物分子刷在药物输送、防污涂层以及锂离子电池等方面的应用。

嵌段共轭聚物的三维螺型结构

何凤[*]

南方科技大学

*通讯作者：hef@mail.sustech.edu.cn

【引言】

聚合物胶束可广泛应用于药物缓释和环境保护。它在制备纳米电子器件和纳米复合材料方面也具有重要的应用前景。由于胶束在选择性溶剂中的稳定性，以聚合物胶束为模板可以制备出许多性能优良的新材料。有序或对称结构的粒子给人以视觉上的美感。无论是纳米粒子还是微粒，粒子的形态对其物理甚至化学性质都有重要影响。因此，制备规则纳米粒子或微粒具有重要的理论研究和实际应用意义。螺型位错是常见的晶体生长机制。晶体通常通过螺旋位错形成三维螺旋形。这种形状在无机晶体（如 MoS_2，$MoSe_2$，WS_2，WSe_2）中比较常见，但在有机晶体中较少见。这种螺旋结构主要在石蜡和聚乙二醇中可观察到。由于其特殊的结构，使得其在催化等领域具有独特的性能。但是，要获得尺寸均匀、结构良好的三维螺旋结构仍然是一个巨大的挑战。尤其是在分子间弱相互作用形成的大分子胶束体系中，这一点尤为困难。

【实验方法】

通过 PPV（聚对苯乙炔）的醛与 PEG（聚乙二醇）链末端的胺基形成席夫碱，再用三乙酰氧基硼氢化钠还原亚胺，合成了两亲性 $PPV_{10}-b-PEG_n$ 嵌段共聚物。利用透射电镜

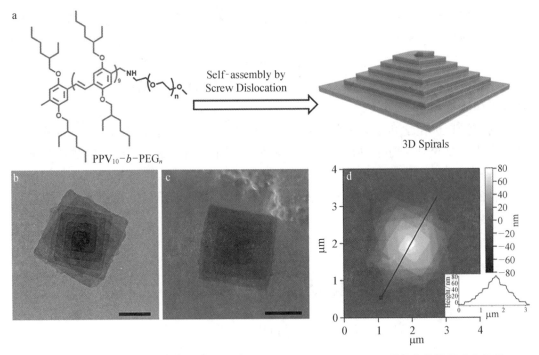

图 1 嵌段共聚物的三维螺旋型结构。(a) 嵌段共聚物 PPV_{10}-b-PEG_n 的螺旋位错驱动自组装形成三维螺旋的化学结构示意图。(b) TEM，(c) HIM 和 (d) 3D 螺旋胶束的 AFM 图像。

(TEM)，氦离子显微镜(HIM)，原子力显微镜(AFM)，紫外吸收光谱仪等对聚合物胶束的形貌和结构进行了探讨。通过开尔文原子力显微镜，导电原子力显微镜对聚合物胶束的电学性能进行了研究。

【结果】

采用自底向上的组装方法制备出三维螺旋嵌段共聚物胶束。通过控制生长溶液的浓度，可以调整所形成胶束的大小，实现从二维正方形到三维螺旋状的形态转变。考虑到在此体系中的低结晶度，我们认为三维螺旋结构是螺旋位错驱动的，嵌段共聚物相对较弱的 π-π 相互作用主导面内堆积。同时，这些三维胶束的表面电势几乎均匀地分布在螺旋的不同位置，且电导率与螺旋层的数量成反比。这表明，由螺旋位错驱动的三维螺旋结构是一种有前景的纳米材料。

Bio-inspired Mechano-functional Gels Through Multi-phase Order-structure Engineering

Liu Mingjie*

Beihang University, Xueyuan Road #37, Haidian District, Beijing, China
* Corresponding author: liumj@buaa.edu.cn

【Introduction】

Adaptive gel materials can greatly change shape and volume in response to diverse stimuli and thus have attracted considerable attention due to their promising applications in soft robots, flexible electronics, and sensors. In biological soft tissues, the dynamic coexistence of opposing components (for example, hydrophilic and oleophilic molecules, organic and inorganic species) is crucial to provide biological materials with complementary functionalities (for example, elasticity, freezing tolerance, and adaptivity). Taking inspiration from nature, we developed a series of high mechanical performance soft active materials, so-called organohydrogels, based on a multiphase synergistic strategy. Traditional techniques such as post-polymerization modification, interpenetrating network, and controlled micro-phase separation are combined with the binary complementary concepts to design and fabricate new organohydrogels with the diverse topology of heteronetworks. Meanwhile, the synergistic effect of heteronetworks provided the organohydrogels with unprecedented mechanical functions such as freeze-tolerance, programmed high-strain shape memory, and shaking insulation. Their applications in anti-biofouling, thin-film fabrication, flexible electronics, and actuators are also explored.

Polysaccharide-based Recoverable Double-network Hydrogel with High Strength and Self-healing Properties

Lei Kun[1], Li Zhao[2], Zhu Dandan[1], Sun Chengyuan[1], Yang Chongchong[1], Zheng Zhen[1], Wang Xinling[1,*]

1 Shanghai Jiao Tong University   2 Beijing Institute of Technology
* Corresponding author: xlwang@sjtu.edu.cn

【Introduction】

Polysaccharide-based hydrogels (PSBHs) have received significant attention for numerous bio-applications due to their biocompatibility and non-immunogenic performance. However, traditional PSBHs show poor mechanical strength, limiting their extensive applications in tissue engineering and bio-devices. There have been some studies on enhancing the mechanical performances; however, in practical application, the PSBHs are usually encumbered by problems like photo-polymerization without in *situ* formation, and non-degradability with the introduction of PAAm or PAAc derivatives. Most reports on polysaccharide-based double network hydrogels mainly focus on the functionality, such as self-healing and NIR responsive abilities. However, few systematic studies have been conducted to evaluate the strength.

【Experimental】

A novel PSBH with superior mechanical and recoverable properties by integrating the synergistic and complementary interactions of oxidized sodium alginate (SA-CHO) gel and agarose (Aga) gel was fabricated by the one-pot method. Mechanical properties, cyclic compression, and swelling properties tests were conducted to evaluate the mechanical properties and stability of PSBHs. FTIR, SEM, and DLS were used to analyze the synergistic interaction mechanism. The rheological and self-healing assays were performed to explore the cyclic shape ability and self-healing.

【Results】

For the SA-CHO/Aga DN hydrogels, the excellent mechanical properties including high strength, modulus, and fracture strain were ascribed to the synergy and complementarity interactions of the soft, mechanically weak SA-CHO network and the rigid, mechanically strong Aga network. The hydrogen bond-based rigid Aga network provided a platform for supporting stress and strengthening the structure, thus greatly enhancing the mechanical strength of PSBHs. The dynamic covalent bond-associated soft SA-CHO network conferred good extensibility.

Due to the thermo-reversible sol-gel transition ability of Aga, the DN gels show recoverable ability. In other words, the broken gel structure could be restored by the transition of sol and gel at a temperature above the melting point.

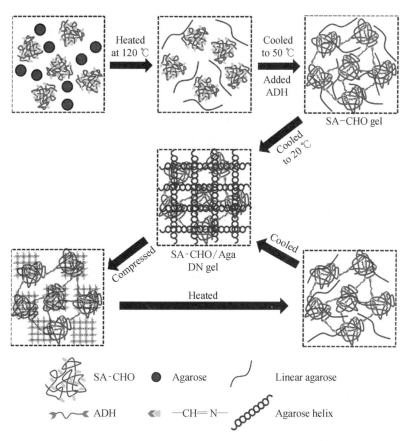

Fig. 1　Fabrication of thermo-reversible and recoverable SA-CHO/Aga DN hydrogels

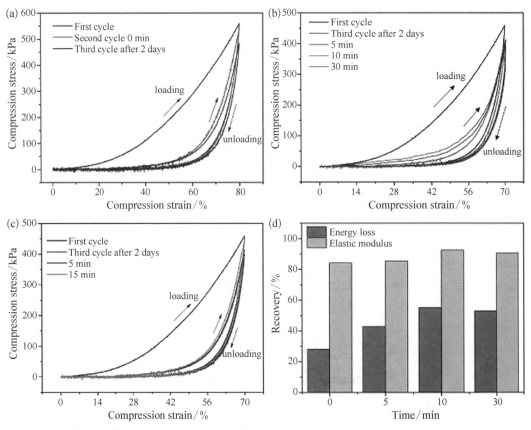

Fig. 2 Loading-unloading cycles of DN gels (a); disposed of (b) at the dynamic acyl hydrazone bond endows the DN gel with 120℃; (c) 50℃. (d) Recovery abilities of DN gel at 120℃ self-healing ability

A Self-healing Hydrogel with Pressure Sensitive Photoluminescence for Remote Force Measurement and Healing Assessment

Li Ming[1,2,*], Li Weijun[2], Xia Zhenhai[3], Eduardo Saiz[1]

1 Imperial College London 2 China University of Petroleum (Beijing) 3 University of North Texas

* Corresponding author: m.li19@imperial.ac.uk

【Introduction】

Self-healing dimming materials exhibit interesting colour changes when exposed to external stimuli such as light, temperature, pH, and force. These interesting properties are important for the advancement of new colour-developing and fluorescent colour-changing materials for applications in sensors, structural health monitoring systems, optical force measuring devices, flexible devices, tissue engineering, and regenerative medicine. Hydrogels are hydrophilic polymers that swell in water but are insoluble in water. These materials have a high water content, good flexibility, and a strong ability to penetrate small molecules. Besides, the unique network structure provides hydrogels with excellent biocompatibility and self-healing abilities. However, in practical applications, hydrogels exhibit relatively long healing times, stimuli dependence, poor healing performance in humid environments, and unstable optical performance. It would be of great interest to prepare the hydrogels with ultra-fast self-healing properties (no external stimuli, multiple environmental adaptations) and stable optical response.

【Experimental】

A fluorescence-responsive self-healing hydrogel with a triple network structure was prepared by gel crystallization, and further explored the self-healing performance of the hydrogel in different media, the fluorescence response performance, as well as the relationship between fluorescence excitation intensity and external force, self-healing efficiency.

【Results】

We proposed a multi-network structure of a fluorescence-responsive self-healing hydrogel (hydrogen bond). The molecular structure of the hydrogel is shown in Fig. 1. The backbone network was formed during the cooling process of the sol-gel in which the chitosan-agarose, chitosan-PVA, and agarose-PVA were strongly crosslinked. The self-healing property of the prepared hydrogels was attributed to a reversible network consisting of the two dynamic covalent cross-linking reactions of PVA-borax-PVA and PVA-glycerol-PVA in the hydrogel. The quantum dots were uniformly dispersed in the hydrogel to ensure stable fluorescence response because of the static hydrogen bond network between PVA and the quantum dots.

This new type of photo fluorescent hydrogel that exhibits excellent self-healing properties in various environments (air, water, petroleum ether, and salt solution). In a gaseous medium, the hydrogel exhibited a 100% tensile strength after 30 seconds of

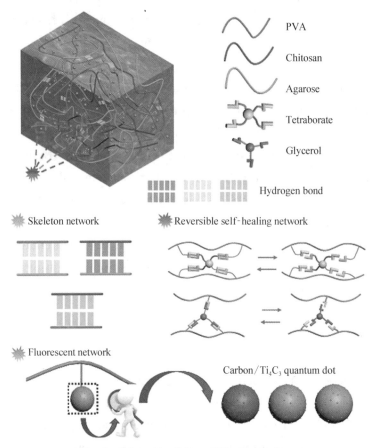

Fig. 1　Schematic of the self-healing hydroge

healing compared to an original, unbroken hydrogel sample. After the damaged area was self-repaired, the same sample area could be rotated at 1 800° without breaking. In a liquid medium, the hydrogel had 90% of its original tensile strength after healing for 60 seconds compared to an unbroken hydrogel sample. The fluorescence performance was shown not to be affected by external conditions (temperature, humidity, and pH). The hydrogel had a pressure remodelling effect. There was no substantial difference in the tensile strength performance and self-healing performance between the reshaped hydrogel obtained after pressing for 10 minutes under an external force of 2 kg and the newly made hydrogel (Fig. 2).

　　The unique relationship between the fluorescence excitation intensity of the hydrogel and the external pressure exerted on it allowed for external pressure measurements to be taken by determining the fluorescence intensity variation (Fig. 3). The self-healing efficiency could be measured using an optical measurement method without damaging the material. As such, the as-developed hydrogels demonstrate a wide variety of potential applications because of their excellent self-healing properties,

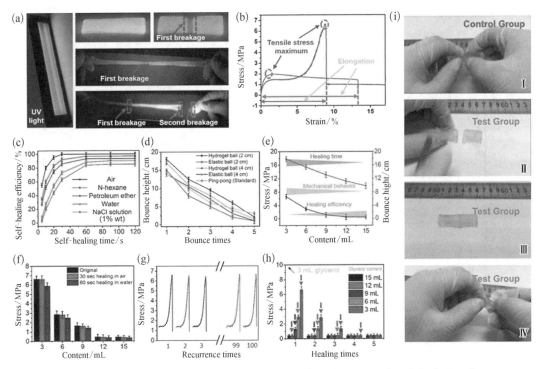

Fig. 2　Excellent fluorescence properties and self-healing properties of the hydrogel

pressure remodelling ability, and fluorescence properties. These hydrogels can be used as light-responsive biosensors, optical force measuring devices, plastic tissue, smart self-healing devices, and biomedical materials.

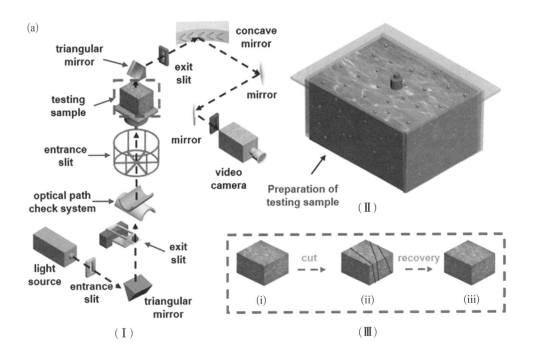

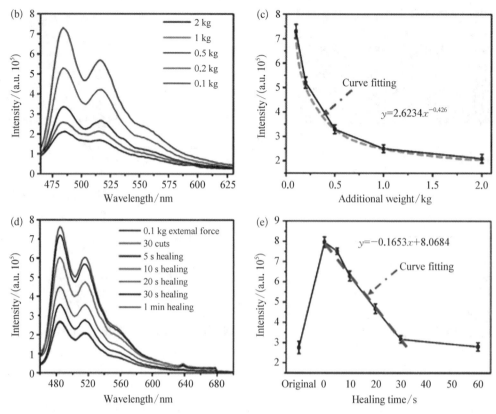

Fig. 3 (a) Mechanism of a hydrogel fluorescence emission intensity test; (b-e) Relationship between the fluorescence intensity of the hydrogel and "its state (original, cut, healed)/the external force to which the hydrogel was subjected"

Preparation of Polyamino Acid Self-Healing Hydrogels Based on 2-Ureido-4[1*H*]-Pyrimidinone

Shi Zhen, Wang Qi, Li Guifei*, Zong Hongjie, Yin Jingbo*

Shanghai University

*Corresponding authors: liguifei0912@163.com; jbyin451@163.com

【Introduction】

Hydrogen bonding is a driving force of great significance in nature life self-assembly. 2-Ureido-4 [1H]-pyrimidinone (UPy), as a self-complementary quadruple hydrogen bonding, is the most widely studied hydrogen-bonding motif, which has a favorable dimerization constant is methylbenzene. Self-healing hydrogels can be devised for stem cell delivery and therapy as an injectable cell carrier, getting more and more attention from researchers. Currently, self-healing hydrogels have been extensively reported, but few reports are focused on the development of self-healing hydrogels using UPy motifs as cross-linking points, especially for tissue engineering applications. The reason is that most self-healing hydrogels employ non-degradable polymers, such as polyethylene glycol (PEG), polyacrylates, dextran, for the substrates. Poly(l-glutamic acid) (PLGA) is a biodegradable and biocompatible polypeptide, which has been successfully used in a variety of tissue reconstructions.

【Experimental】

UPy moieties were grafted on the PLGA backbone, employing α-hydroxy-ω-amino poly(ethylene oxide) (HAPEO) as the connection agent. Self-healing hydrogels were developed through UPy units as crosslinked blocks. Self-healing and shear-thinning behaviors of the hydrogels were performed at 37℃ by oscillatory mode on a rheometer. The injectable test was verified with an 18G needle (outside diameter=1.20 mm, injection speed=1 mL/min).

【Results】

The reversible nature of UPy dimers endows the hydrogel with characteristics of self-healing and shear-thinning properties as well as the cross-linked network. At the microscopic level, when the strain applied increases, the hydrogel's cross-linked network begins to break down. When the strain reaches a certain degree, the gel network is completely broken down and transformed into a sol. In the strain-modulus curve, as shown in Fig. 2(a), when the strain was 200%, the storage modulus G' was the same as the loss modulus G'', where was the gel-sol conversion point. The Cyclic alternate strain test was conducted on the hydrogels, as shown in Fig. 2(b), indicating that the hydrogels had good self-healing performance. Through the determination of the viscosity-shear rate, as shown in Fig. 2(c), showing the characteristics of shear-thinning. At high-speed shear, the crosslinking network was damaged, with the corresponding viscosity decreased. When the shear force was removed, the UPy dimers

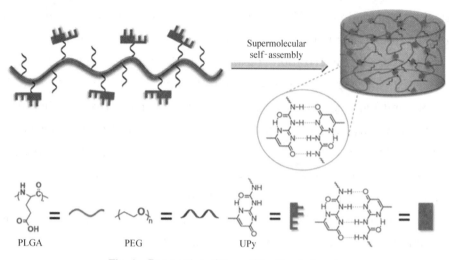

Fig. 1　Preparation of the self-healing hydrogels

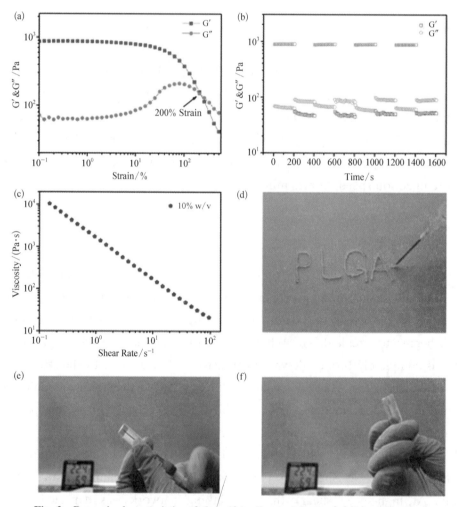

Fig. 2　Dynamic characteristics of the self-healing hydrogels: (a)(b) self-healing ability; (c) shear-shinning ability; (d)(e)(f) injectability

were recombined to form a new crosslink point and realize the healing of the hydrogels. The shear-thinning property of hydrogels made it convenient to deliver from outside to the inside of the vials during injection (Fig. 2d – f).

All the dynamic characteristics of the self-healing hydrogels showed the great potential in tissue engineering application.

Injectable Simvastatin-loaded Micelle/Hydrogel Composites for Bone Tissue Engineering

Wu Nianqi, You Jiahui, Yan Shifeng*, Yin Jingbo*

Shanghai University

* Corresponding authors: yansf@staff.shu.edu.cn; jbyin@oa.shu.edu.cn

【Introduction】

Injectable hydrogels have shown broad application prospects in bone tissue engineering because of minimally invasive surgical procedure, in situ gelations, filling of complex shape defects, and simulation of the natural extracellular matrix. Simvastatin (SIM) is a commonly used lipid-lowering drug with the advantages of safety, stability, and low costing shows great potential in promoting bone formation. Moreover, SIM can also provide additional functions like promoting neovascularization and possessing anti-inflammation property. However, the solubility of SIM is pretty poor in water. Besides burst release of the drug is also a common problem when SIM is loaded directly into the material matrix, which greatly impacts the maintaining of effect concentration. To homogeneously disperse SIM and achieve sustained release of SIM, we designed a kind of injectable micelle/hydrogel composites based on maltodextrin and carboxymethyl chitosan.

【Experimental】

The composite hydrogels were fabricated by self-crosslinking between aldehyde-modified MDex-C16 micelles (micelle-CHO)-loaded oxidized maltodextrin (OMDex) solution with carboxymethyl chitosan (CMCS) solution. The preparation of micelle/hydrogel composites, drug loading, and release was investigated. The mouse osteogenic precursor cells (MC3T3-E1) were encapsulated within the composite hydrogels to evaluate their cytocompatibility and osteogenic capability (Fig. 1).

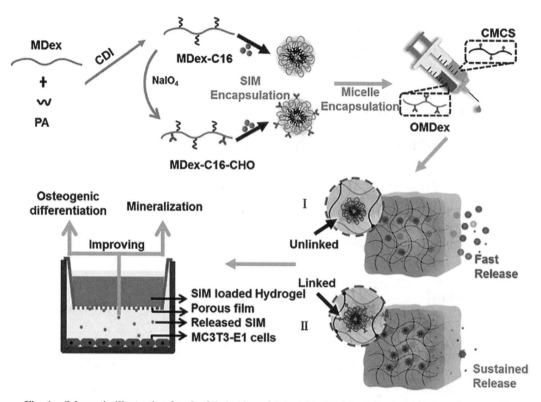

Fig. 1 Schematic illustration for the fabrication of injectable SIM-loaded micelle/hydrogel composites

【Results】

Quantitative analysis of alizarin red staining and alkaline phosphatase (ALP) activity assay of MC3T3-E1 cells cultured on the hydrogels were conducted to verify the in vitro osteogenic activity of the different composite hydrogels. From the quantitative analysis of alizarin red staining, the MC3T3-E1 cells exhibited a continuously increasing mineralization after 7 and 14 days of culture. The highest mineralization degree was found in the SIM/Micelle-CHO hydrogel group. As shown in Fig. 2, ALP activity was also increased with culture time and achieved the highest value on the SIM/Micelle-CHO hydrogels. These results suggested that SIM/Micelle-CHO composite hydrogel possessed great potential as an injectable system for bone tissue engineering.

Acknowledgements: The work was supported by the National Natural Science

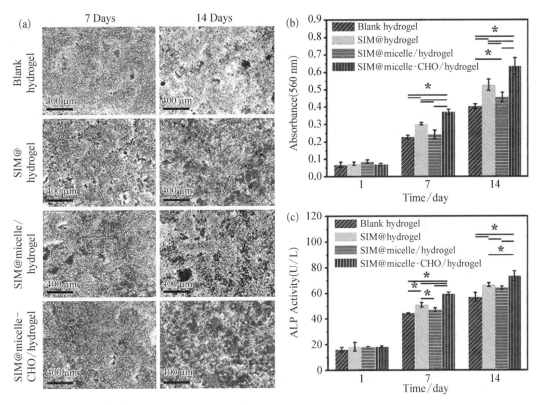

Fig. 2 Osteogenesis of MC3T3-E1 cells. (a) Alizarin red staining at 7 and 14 days; Quantitative results of (b) alizarin red staining and (c) ALP staining

Foundation of China (No. 51873101).

Symbiont Mutagenesis and Characterization as a Potential Method to Alter Holobiont Stress Tolerance: A Study on Hydra Viridissima

Ye Siao, Meenakshi Bhattacharjee, Evan Siemann

Rice University

* Corresponding author: thomasyesiao@gmail.com

【Introduction】

The hologenome theory proposes that changes in symbiont could alter also holobiont traits, thus enables symbiotic organisms to adapt rapidly and persist in stressful environments. New symbiont acquisition or change in symbiont frequencies can improve holobiont stress tolerance, however, whether symbiont mutations can be induced and selected to alleviate holobiont stress has not been demonstrated. Compared to direct manipulation on host genes, modifications on the microbial symbionts could be faster, easier, and transferable. Confirming the feasibility of such symbiont-mediated host phenotype alteration would shed light on ecological conservation and agriculture practice.

【Experimental】

To explore whether we could alter holobionts' stress tolerance by symbiont mutagenesis and selection, we used green hydra (*Hydra viridissima*) and endosymbiotic algae (*Chlorella variabilis* NC64A) together with UV tolerance as our model system. We used UV-C (200 – 280 nm) to induce mutations in the algae. We used UV-B (300 – 315 nm) as the selection factor in our experiments because of its detrimental effects on aquatic invertebrates), which can be attenuated by endosymbiotic algae via shading or mycosporine-like amino acids (MAAs). Multiple algals replicated were mutated independently and then selected, and controls were maintained for each replicate. Algal UV-B tolerance was tested in vitro and then tested in vivo by injecting into the non-symbiotic hydra respectively. UV－B tolerance of the hydra and the algae were then analyzed and compared.

【Results】

Symbiont mutagenesis in vitro altered their UV-B resistance as well as that of holobionts receiving mutated algae. Also, hydra UV-B tolerance was positively correlated to that of the algae they were hosting, as the hydra associated with UV-B tolerant algal strains exhibited higher UV-B resistance. However, chronic low-level UV-B selection decreased algal resistance to acute high-level UV-B and did not affect UV-B resistance of holobionts. The variations in algal UV-B tolerance and hydra UV-B tolerance were largely due to mutagenesis rather than selection. Our results suggest symbiont mutagenesis and trait-based identification may be more effective than assisted evolution in holobiont phenotype alteration. Symbiont mutagenesis potentially provides a new approach to achieve desired traits in holobionts, which may have agricultural, forestry, and conservation applications.

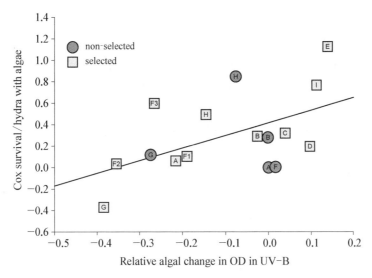

Fig. 1 The correlation (line) between algal change in OD in UV-B (Fig 2B) and Cox survival coefficient for different non-selected (dark circles) or selected (light squares) populations of algae

Mussel-inspired Adhesive, Self-healing, and Injectable Poly (L-glutamic acid)/ Alginate-based Hydrogels

You Jiahui, Wu Nianqi, Yan Shifeng*, Yin Jingbo*

Shanghai University

*Corresponding authors: yansf@staff.shu.edu.cn; jbyin@oa.shu.edu.cn

【Introduction】

Injectable hydrogels have aroused much attention for the advantages such as minimally invasive surgery, avoidance of surgical trauma, and filling and repairing irregularly shaped tissue defects. Mussel-inspired injectable hydrogels can be immobilized on the surface of tissues, resulting in stable biomaterial-tissue integration. However, the commonly used biomimetic mussel-inspired hydrogels are prepared by the oxidation of catechol groups, which involves the introduction or production of cytotoxic substances. Moreover, mussel-inspired hydrogels generally display weak mechanical strength and poor adhesiveness because of the consumption of catechol groups during oxidation. Herein, we described a strategy to prepare mussel-inspired injectable hydrogels via the Schiff base reaction. A series of injectable mussel-inspired adhesive, self-healing hydrogels were fabricated by in situ crosslinking of hydrazide-modified poly (L-glutamic acid) (PLGA-ADH) and dual-functionalized alginate (catechol-and aldehyde-modified alginate, ALG-CHO-Catechol).

【Experimental】

^{1}H NMR spectra and UV-vis spectra were characterized to confirm the conjugation of catechol groups onto ALG-CHO. Rheological tests and adhesion measurements were operated to evaluate the mechanical and adhesive properties of the obtained hydrogels. The in vivo hemostatic ability of the hydrogels was investigated using a rat hemorrhaging liver model (male SD rats, 150 – 200 g).

【Results】

ALG-CHO-Catechol conjugates with controllable catechol grafting ratios and hydrophilicity were synthesized. The mussel-inspired adhesive, self-healing, and injectable PLGA/ALG-CHO-Catechol hydrogels were prepared via self-crosslinking of PLGA-ADH and ALG-CHO-Catechol by Schiff base reaction, which avoided the introduction of small molecular oxidants and preserved the catechol functional groups (Fig.1). The gelation occurred in a reasonable time. Compared with the oxidized ALG-

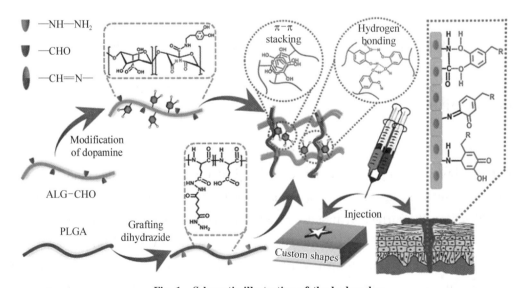

Fig. 1　Schematic illustration of the hydrogels

CHO-Catechol hydrogels, the PLGA/ALG-CHO-Catechol hydrogels showed greatly improved mechanical properties and adhesive properties. The PLGA/ALG-CHO-Catechol hydrogels also exhibited strong self-healing ability and good biocompatibility with adipose stem cells. In vivo antibleeding displayed superior hemostatic capacity of the PLGA/ALG-CHO-Catechol hydrogels. The injectable PLGA/ALG-CHO-Catechol hydrogel system demonstrates attractive properties and has shown promise as a suitable hemostatic agent with high performance (Fig. 2).

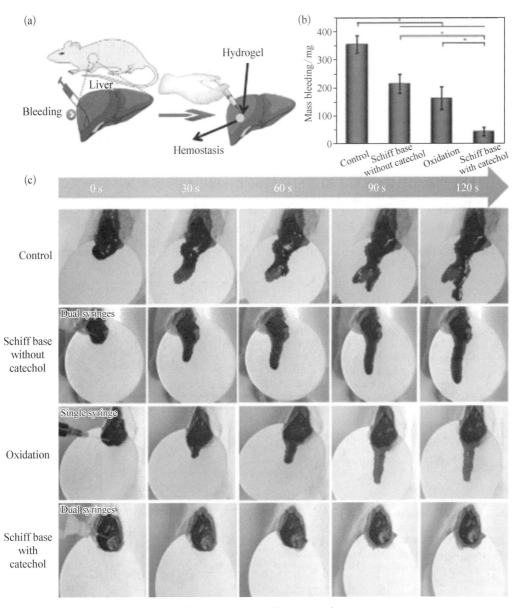

Fig. 2　Hemostatic capability of the hydrogels

高分子水凝胶整合酶的生物医学应用

王启刚 *

同济大学化学科学与工程学院

* 通讯作者：wangqg66@tongji.edu.cn

  酶是细胞代谢的基础，它通过复杂有序的生物化学反应来调节细胞物质转换、能量生成及免疫防御等生命过程。酶离开细胞保护后通常活性下降且无法进行高效有序连续反应，合成高分子水凝胶因类细胞外基质的三维网络是酶与多酶固定化优异载体。本研究中通过多酶催化聚合及定向组装来构建高分子水凝胶整合酶，借助其类有氧代谢功能，实现其在疾病快速诊断和生物治疗领域的应用。首先通过酶复合体的级联催化聚合与可控链增长等仿生高分子合成新方法，构建了超分子与高分子复合水凝胶来定向包埋多酶并维持最佳生物催化条件，克服了酶蛋白在固定化过程中结构破坏及活性损失。然后通过高分子水凝胶整合酶来构建生物诊断体系，借助类有氧代谢过程的电子转移和信号传导，实现疾病代谢物快速物理检测。最后通过高分子水凝胶整合多酶干预活性氧的代谢反应并破坏肿瘤细胞脆弱的氧化还原平衡，形成了基于多酶催化生物治疗的新策略。

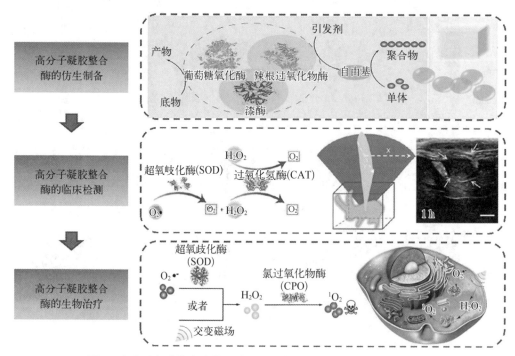

图 1 高分子凝胶整合酶体系的仿生制备及临床检测和生物治疗应用

仿生黏附可控界面

王树涛*

中国科学院理化技术研究所仿生智能界面科学实验室，北京，100190

*通讯作者：stwang@mail.ipc.ac.cn

生物界面黏附是界面化学研究中的前沿热点之一，它不仅有助于我们认识生命的奥秘，还对发展新型功能界面材料及相关技术有着重要意义。向自然学习，为发展新材料源源不断地提供新原理和新思路。近几年来，我们研究了几种生物界面上的特殊黏附现象，并受此启发发展了系列仿生黏附可控界面。一是揭示了鸟类羽毛耐撕裂的性能是源于羽毛上的钩—槽—钩机械级联互锁结构，它颠覆了传统认为的简单钩—槽互锁结构，发展了仿蜻蜓干态黏附材料，解决了传统仿苍耳锦纶粘扣易坏、噪声大等问题；发展了仿肾小管内壁的抗矿物黏附界面材料。二是提出结构匹配和分子识别的协同仿免疫界面识别理念，利用化学刻蚀、气相沉积、电化学沉积、模板复形、电纺等技术构筑了系列仿免疫CTC捕获芯片；提出界面乳液聚合方法，构筑了系列形状可控（从Janus到多孔）与表面化学可控的仿免疫磁珠微球。三是揭示了伤口血痂的微观结构，构筑了系列仿血痂伤口辅料，有效地促进了伤口愈合。

单分子层丝素纳米带及全丝素基纳米摩擦发电机

牛欠欠，范苏娜，张耀鹏*

东华大学

*通讯作者：zyp@dhu.edu.cn

【引言】

随着电子器件的发展，具有优异生物相容性的全降解可植入型器件已成为医学领域的研究热点。大多数植入器件仍旧需要额外电池供能。纳米摩擦发电机（TENG）可以在实时监测的同时，将自身机械能转化成电能。为了避免二次手术伤害，可以体内降解的TENG研究十分必要。丝素蛋白是一种力学性能优异、生物相容性好、环境友好的天然高分子材料,广泛应用于生物医药等领域。丝素基材料的性能主要受其分子量、β-折叠微晶结构和微纤结构的影响。其中,丝素纳米纤维(SNF)作为蚕丝的基本构筑单元,保留了天然丝的微观结构。本文采用次氯酸钠/TEMPO/溴化钠体系溶解丝素,制备单分子层厚的丝素纳米带(SNR)。采用 SNR 膜(SNRF)与再生丝素蛋白膜(RSFF)作为摩擦层,制备全丝素 TENG。

【实验方法】

将脱胶丝、去离子水、TEMPO、溴化钠按一定质量比(1∶90∶0.02∶0.1)混合,再加入 0.016 mol 次氯酸钠。反应过程中,通过添加氢氧化钠水溶液使反应体系的 pH 维持在 10~10.5 之间;反应结束后将其水洗、超声、离心,最终得到 SNR 悬浮液。采用真空抽滤法,制备 SNRF。采用 9M 溴化锂水溶液溶解脱胶丝,再经过透析、浓缩,制备浓度为 20 wt%的再生丝素蛋白水溶液(RSF);采用浇铸法制备 RSF 膜(RSFF)。通过在膜表面蒸镀导电层镁,连接导线,制备 SNRF/RSFF TENG。包裹层为经过一定后处理的 RSFF。

【结果】

为了探究悬浮液中 SNR 的尺寸,我们采用 TEM 和 AFM 对在铜网或者云母片上的干态样品进行表征。图 1(a)和 1(b)分别为 SNR 的 TEM 图和 AFM 图,从图中可以看出 SNRs。图 1(c)是依据 AFM 图像测量 100 根 SNRs 得到的 SNR 宽度分布图。SNR 的宽度为 14~28 nm,平均宽度为 20.2 nm。图 3(d)为 AFM 轻敲模式测得的 SNR 厚度图,SNR 厚度约为 0.38±0.01 nm。可见,SNR 为片层状,具有与 β 片层相似的厚度,为单分子层 SNR。

对 SNRF/RSFF TENG 施加外力,检测输出性能。图 2(a)是 TENG 在外接电阻为 100 MΩ 时的输出电压。图 2(b)为 TENG 在不同外接电阻下的输出性能,其中,最高输出电压为 41.64 V,最大输出能量密度为 86.7 mW/m²。图 2(c)对比了 SNRF/RSFF

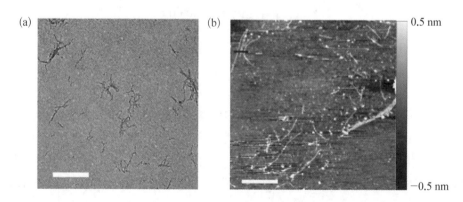

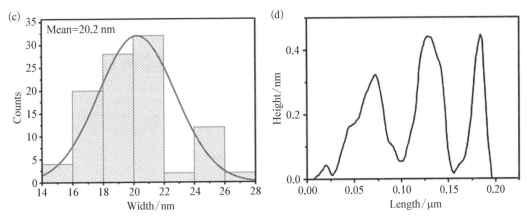

图 1 核糖核酸的形态和大小。SNRS 的 TEM 图像(a)和 AFM 图像(b)，基于原子力显微镜测量的 SNR 宽度分布(c)和高度分布(d)，比例尺为 500 nm

TENG 与其他生物基 TENG 的性能。与其他可植入全降解生物基 TENG(SP：合成聚合物；NP：天然聚合物)相比，SNRF/RSFF TENG 具有更加优异的输出性能。并且 SNRF/RSFF TENG 具有较高的灵敏性，可以监测脉搏跳动(图 2d)。这种 TENG 有望应用于生物医学、传感等领域。

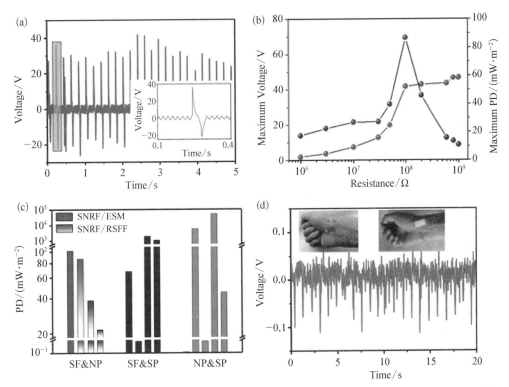

图 2 TENG 在各种输入条件下的输出性能。(a) 频率为 3H2 以下的 TENG 输出电压。(b) 输出电压和 PD 对不同的外部电阻的依赖关系。(c) SNRF/RSFF TENG 的 PD 与以往报道的其他基于生物的 TENGS 的比较。(具有变形式的柱子表示 Mg 导电层，单一纯色形式的柱子表示 Al, ITO 或 Gu 导电层。以往数据取自参考文献)(d) 实时监测手腕脉搏

液晶弹性体基双向形状记忆材料研究

陆海峰,左波,刘莉,王猛,杨洪*

东南大学
* 通讯作者:yangh@seu.edu.cn

【引言】

液晶弹性体是一种典型的双向形状记忆材料,具有形变大、形变可逆等技术优点,在仿生器件、软机器人等领域有着广阔的应用前景。然而,经过长达四十多年的发展,液晶弹性体的研究仍停留在实验室层面,尚未实现工业化应用;限制其应用的关键科学问题是:液晶弹性体在形变过程中产生的应力太小,无法满足实际应用场景的力学性能需求。例如,人的骨骼肌收缩应变大于40%,应力大于0.35 MPa,弹性模量大于10 MPa;对于传统的液晶弹性体材料来说,收缩应变不是问题,应力也勉强可以达到,但是弹性模量一般在0.1~1 MPa之间,低了整整1~2个数量级。本报告从材料科学角度,分析了液晶弹性体基人工肌肉研究的现状与发展趋势,探讨了可能涉及的相关科学问题及研究思路。

【实验方法】

采用将聚氨酯液晶弹性体与聚丙烯酸酯液晶热固体的小分子前体组分混合,再同步交联的技术途径,制备了一种聚氨酯/聚丙烯酸酯互穿网络结构液晶弹性体材料。

【结果】

该聚氨酯/聚丙烯酸酯互穿网络结构液晶弹性体材料的收缩应变、应力、弹性模量分别达到了46%、2.53 MPa和10.4 MPa,首次全面满足了液晶弹性体基人造肌肉的力学性能要求。如图1所示,这种新型人造肌肉材料具有超强的力学性能,当它发生伸缩形变时,其抗拉强度和弹性模量都远远超出了以往的研究结果;在热刺激条件下,该材料可以拉起自身重量30 000倍的物体。

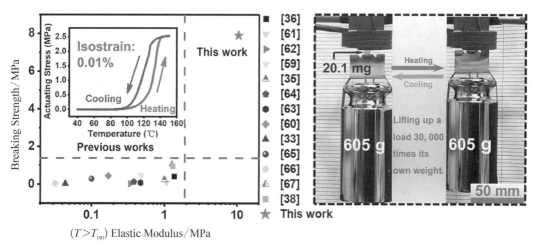

图 1 具有超强机械性能的互穿液晶聚氨酯/聚丙烯酸酯弹性体

纳米分子伴侣调控胰岛素的递送

李畅[1,2,*]，史林启[2]

1 同济大学 2 南开大学

* 通讯作者：lichang@tongji.edu.cn

【引言】

糖尿病是一种以高血糖为特征的代谢性疾病，而胰岛素是目前治疗糖尿病的主要药物之一。胰岛素是一种蛋白质药物，其在储存或皮下注射时，自身很容易发生错误聚集而丧失活性。此外，经皮下注射进入体内的胰岛素很容易被体内的蛋白酶水解，从而影响其治疗效果。作为一类天然分子伴侣，热休克蛋白是维持体内蛋白稳态的重要组成部分，其疏水空腔可以与蛋白质相互作用，既能有效地控制蛋白质的错误折叠与聚集，又能辅助蛋白质的折叠、转运和清除，从而保证细胞的正常生理功能。

【实验方法】

通过自组装获得复合壳层胶束，并模拟热休克蛋白的结构和功能，构建纳米分子伴侣。纳米分子伴侣表面由塌缩的葡萄糖响应性疏水微区与伸展的亲水聚合物链段组成，从而形成类似热休克蛋白的疏水空腔，有效负载和保护胰岛素，抑制其错误聚集并减弱其被蛋白酶降解的程度，从而实现胰岛素的递送及葡萄糖响应性释放，如图 1 所示。利用石

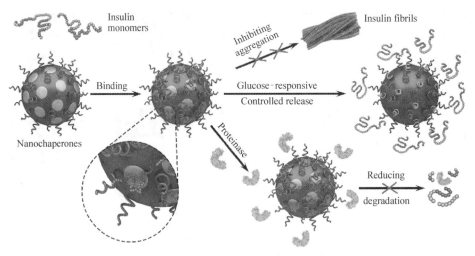

图1 纳米分子伴侣负载、保护和递送胰岛素示意图

英晶体微天平测试(QCM)、荧光共振能量转移试验(FRET)和透射电子显微镜(TEM)等手段研究纳米分子伴侣的保护效果。

【结果】

首先研究了纳米分子伴侣在体外对胰岛素单体的负载及抑制其错误聚集的效果。将纳米分子伴侣与胰岛素单体在 37 ℃共孵育 7 d 后,TEM 结果显示,相较于对照组,纳米分子伴侣显著抑制了胰岛素聚集体的形成(图 2),证明纳米分子伴侣可有效负载并保护胰

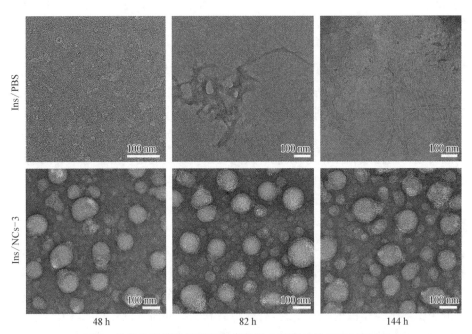

图2 纳米分子伴侣保护胰岛素并阻止其聚集的 TEM 结果图

岛素单体,从而有效抑制其错误聚集。

体内降血糖试验结果表明,在纳米分子伴侣的保护下,胰岛素在37℃孵育7 d或70℃孵育1 h后仍具有明显的降血糖效果,如图3中(a)和(b)所示。另外,负载有胰岛素的纳米分子伴侣与蛋白酶共孵育后,其也具有一定的降血糖效果,葡萄糖耐受试验结果表明,负载有胰岛素的纳米分子伴侣具有明显的体内葡萄糖响应性,如图3中(c)和(d)所示。同时,纳米分子伴侣可有效提高胰岛素的长循环时间,利用其对糖尿病模型小鼠进行持续治疗,模型小鼠的体重和饮水量也有了积极的变化。以上体外和体内试验的结果均表明,纳米分子伴侣可以显著抑制胰岛素单体的聚集,降低胰岛素被体内蛋白酶降解的程度,有效递送胰岛素并根据体内血糖变化释放胰岛素,从而精确调控糖尿病模型小鼠的血糖水平。纳米分子伴侣有望为糖尿病的治疗提供一种新的策略和思路。

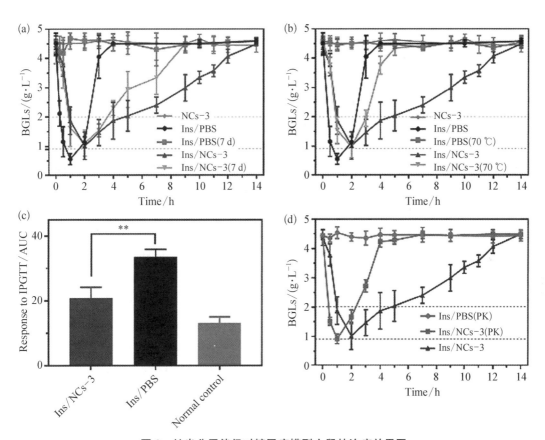

图3 纳米分子伴侣对糖尿病模型小鼠的治疗效果图

SAXS Studies of the Thermally-induced Fusion of Diblock Copolymer Spheres: Formation of Hybrid Nanoparticles of Intermediate Size and Shape

Erik Jan Cornel*

Tongji University, 4800 Caoan Road, Shanghai 201804, China
* Corresponding author: 20310048@tongji.edu.cn

【Introduction】

Dilute dispersions of poly(lauryl methacrylate)—poly(benzyl methacrylate) (PLMA-PBzMA) diblock copolymer spheres (a.k.a. micelles) of differing mean particle diameter were mixed and thermally annealed at 150℃ to produce spherical nanoparticles of intermediate size. The two initial dispersions were prepared via reversible addition-fragmentation chain transfer (RAFT) dispersion polymerization of benzyl methacrylate in n-dodecane at 90℃. Systematic variation of the mean degree of polymerization of the core-forming PBzMA block enabled control over the mean particle diameter: small-angle X-ray scattering (SAXS) analysis indicated that $PLMA_{39}$-$PBzMA_{97}$ and $PLMA_{39}$-$PBzMA_{294}$ formed well-defined, non-interacting spheres at 25℃ with core diameters of 21 ± 2 nm and 48 ± 5 nm, respectively. When heated separately, both types of nanoparticles regained their original dimensions during a 25–150–25℃ thermal cycle. However, the cores of the smaller nanoparticles became appreciably solvated when annealed at 150℃, whereas the larger nanoparticles remained virtually non-solvated at this temperature. Moreover, heating caused a significant reduction in the mean aggregation number for the $PLMA_{39}$-$PBzMA_{97}$ nanoparticles, suggesting their partial dissociation at 150℃. Binary mixtures of $PLMA_{39}$-$PBzMA_{97}$ and $PLMA_{39}$-$PBzMA_{294}$ nanoparticles were then studied over a wide range of compositions. For example, annealing a 1.0% w/w equivolume binary mixture led to the formation of a single population of spheres of intermediate mean diameter (36 ± 4 nm). Thus we hypothesize that the individual $PLMA_{39}$-$PBzMA_{97}$ chains interact with the larger $PLMA_{39}$-$PBzMA_{294}$ nanoparticles to form the hybrid nanoparticles. Time-resolved SAXS studies confirm that the evolution in copolymer morphology occurs on relatively short time scales (within 20 min at 150 ℃) and involves weakly anisotropic intermediate species. Moreover, weakly anisotropic nanoparticles can be obtained as a final copolymer morphology over a restricted range of compositions (e.g. for $PLMA_{39}$-$PBzMA_{97}$ volume fractions of 0.20 – 0.35) when heating dilutes dispersions of such binary nanoparticle mixtures up to 150℃. A mechanism involving both chain expulsion/insertion and micelle fusion/fission is proposed to account for these unexpected observations.

【Experimental and Results】

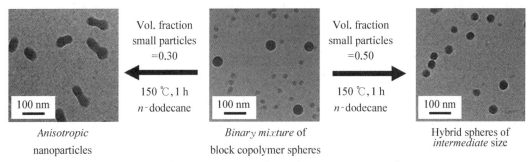

Fig. 1 Dilute dispersions of poly(lauryl methacrylate)-poly(benzyl methacrylate) diblock copolymer spheres of differing mean diameter are mixed and thermally annealed at 150℃ to produce either spherical or non-spherical nanoparticles of intermediate size

Molecular Pillar Approach to Construct 3D Nanomaterials for Energy Storage

Sun Jinhua[1,2,*]

1 Chalmers University of Technology 2 Umeå University

* Corresponding author: jinhua.sun@outlook.com

【Introduction】

Two dimensional (2D) nanomaterials are considered to be promising for applications in almost every important energy-related fields. The ultra-high surface area and porous structure of graphene-related materials (GRMs) are efficient for gas storage (e. g., hydrogen) and can accommodate large amounts of charges for electrochemical energy storage. In this regard, their performance could be further boosted by integrating with other functional materials which could provide added values. This could be realized with the strategy of constructing a 3D structure upon the 2D nanosheets based on the rich surface chemistry of 2D GRMs.

【Experimental】

Based on our understanding of the interlayer structure of 2D materials, a molecular pillar approach was developed to construct hybrid 2D-2D materials, which composed of the perpendicularly oriented covalent organic framework (COFs) and graphene. The thickness and density of COF nanosheets on the surface of graphene were tubed by varying the ratio between graphene and pillar molecules. The edge-on the orientation feature of COF on graphene preserves after carbonization under the assist of molten salt. The nanostructure and morphology of both COF-graphene and carbonized COF-graphene were investigated by SEM, TEM, BET, XRD, Raman, and XPS. The electrochemical performance of carbonized COF-graphene for supercapacitors was characterized.

【Results】

In a controllable two-step synthesis method, the Benzene-1, 4-diboronic acid (DBA) molecules were first covalently attached to graphene oxide (GO) with high density, and then used as nucleation sites for directing vertical growth of COF-1 nanosheets (v-COF-GO). The edge-on anchoring of DBA molecules to GO determines the formation of COF-1 nanosheets forest in edge-on orientation relative to GO (Fig.1). The thickness of COF-1 nanosheets was precisely controlled from ～ 3 to

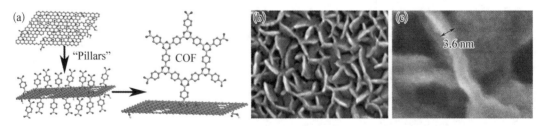

Fig. 1 (a) Growth of vertical COF-1 nanosheets using DBA as molecular nucleation sites grafted on GO. (b) SEM and (c) STEM images of v-COF-GO

15 nm (several to tens layers of COF-1) by varying the loading of DBA. However, the same reaction performed in the absence of molecular pillars resulted in the uncontrollable growth of thick COF-1 platelets parallel to the surface of GO.

The hybrid COF-GO material was successfully converted into a highly conductive boron-doped carbon phase under the protection of molten salts. The obtained carbon materials preserved the nanostructure of the precursor with ultrathin porous carbon nanosheets grafted to reduced GO in edge-on orientation. Owing to the unique nanostructure of hybrid 2D-2D carbon materials which promised the fast electron transfer from vertical porous carbon nanosheets to highly conductive reduced GO, the carbonized v-COF-GO acted as electrode materials exhibited high electrochemical performance for supercapacitors (Fig. 2).

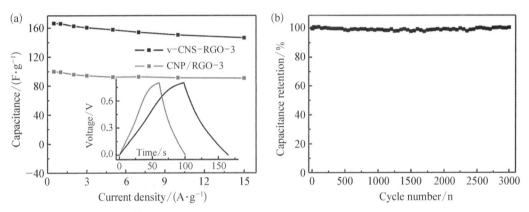

Fig. 2 (a) Capacitance for carbonized samples with vertical (v-CNS-RGO-3) and parallel (CNP/RGO-3) orientation of nanosheets; (b) Cycling performance of the v-CNS-RGO-3 electrode

Confinement of Single Polyoxometalate Clusters in Molecular-scale Cages for Improved Flexible Solid-state Supercapacitors

Wang Meiling*, Liu Xuguang

Taiyuan University of Technology

*Corresponding author: wmlsdym@163.com

Here we realize supramolecular confinement of single polyoxometalate (POM)

cluster precisely in polypyrrole (PPy) hydrogel-wrapped CNT framework with molecular-scale cages. Such a hybrid hydrogel framework demonstrates ultra-high loading (67.5 wt%) and extremely uniform dispersion of individuals of $H_3[P(Mo_3O_{10})_4]$ (PMo_{12}) molecules as demonstrated by subångström-resolution HAADF-STEM. Therefore it exhibits better supercapacitor performance compared to conventional composite systems. The flexible solid-state supercapacitor shows a high energy density of 67.5 $\mu Wh \cdot cm^{-2}$ at a power density of 700 $\mu W \cdot cm^{-2}$ and delivers high capacitance retention of 85.7% after 3 000 cycles. Moreover, the flexible device exhibits excellent mechanical stability. Density functional theory calculations reveal that the wrapped "fishnet-like" hydrogel creates a cage structure of 1.8 nm for precise storage of PMo_{12} molecule (diameter = 1.05 nm), therefore lead to the mono-dispersion of single PMo_{12} molecules on the hybrid hydrogel. The "caging" effect also activates the PMo_{12} molecule to enhance its charging/discharging performance by introducing new reactive sites for proton transfer. We believe the designing of suitable cage structure can be used for the construction of other POM-based hybrid hydrogels, thereby achieving their mono-dispersity and performance enhancement.

Fig. 1 Confinement of single polyoxometalate clusters in molecular-scale cages for flexible solid-state supercapacitor

Lactose Targeting Photodynamic Antibacterial Materials for Pseudomonas Killed

Zhu Yiwen, Yu Bingran*, Fujian Xu*

Key Laboratory of Chemical Resource Engineering, Beijing University
of Chemical Technology (BUCT), Beijing 100029, China
* Corresponding authors: xufj@mail.buct.edu.cn;
yubr@mail.buct.edu.cn

【Introduction】

P. aeruginosa is a threatening pathogen in ophthalmic diseases. Once a Pseudomonas aeruginosa biofilm is formed, antibiotics lose their effectiveness because they cannot penetrate the biofilm. Photodynamics has received widespread attention owing to target selectivity, remote controllability and favorable biosafety in normal tissues. ROS can not penetrate through the robust resistance of Gram-negative bacteria outer membranes relatively porous cell walls of Gram-positive bacteria, thus, photosensitizer can not kill Gram-positive bacteria upon light irradiation. The α-D-galactose can specifically bind to Lec A lectin on the surface of Gram-negative bacteria P. aeruginosa. To solve the biofilm of P. aeruginosa in the cornea, we proposed the α-D-galactose-targeted photodynamic antibacterial strategy.

【Experimental】

Firstly, a double bond is introduced into the α-D-galactose, and then acid red 94 is introduced through RAFT polymerization and amino-epoxy ring-opening reaction. The α-D-galactose-targeted photodynamic antibacterial material can be combined with P. aeruginosa pass through the α-D-galactose. And then, acid red 94 produced ROS upon light irradiation, leading to the death of the bacterial cell.

【Results】

The α-D-galactose-targeted photodynamic antibacterial material shows excellent targeting properties, higher antibacterial activity, and low cytotoxicity. Based on the results of the in vitro experiments, we conducted a rabbit cornea infection experiment. As a result, our material has a good therapeutic effect on rabbit cornea P. aeruginosa infection. Therefore, our lactose-targeted photodynamic antibacterial strategy against P. aeruginosa infection shows excellent prospects.

Self-assembled Nucleotide/Saccharide-tethering Polycation-based Nanoparticle for Targeted Tumor Therapy

Yu Dan, Liu Siyuan, Yu Bingran*, Wang Zhengang*, Xu Fujian*

Beijing University of Chemical Technology
* Corresponding authors: xufj@mail.buct.edu.cn;
yubr@mail.buct.edu.cn; wangzg@mail.buct.edu.cn

【Introduction】

Hydrophobic drugs, as a vast majority of the chemotherapeutics, have suffered from low biosafety, bioavailability, and the anticancer effect, due to poor solubility and short half-life, and low specificity. Supramolecular systems derived from endogenous materials have been considered promising for the fabrication of drug vehicles, due to their full biocompatibility. A variety of self-assembled DNA-based drug vectors, which loaded the anticancer drugs through intercalation interactions, have been fabricated to deliver drugs *in vivo*. However, the specific interactions between the carrier materials and the drug molecules limited the type of chemotherapeutics. Hence, the development of novel concepts to construct drug vehicles for efficient encapsulation of hydrophobic drugs and targeted therapy is highly desirable to biomedical science. It remains a challenge to create an effective noncovalent linkage between the components within the drug vectors. In our work, we report a noncovalent concept to construct drug vectors, based on self-assembly of the tumor-targeting cationic P(GEA-*co*-Fru) with self-associating GMP and a hydrophobic nucleoside analogue drug clofarabine, for the treatment of breast tumors.

【Experiment】

A endogenous self-associating nucleotide (guanosine monophosphate) noncovalently bonded with a hydrophobic nucleoside analogue drug (clofarabine) to form nanofibrils, which were transformed into spherical nanoparticles by assembling with a fructose/ethanolamine-functionalized star-like poly(glycidyl methacrylate)-based cationic polymer [named P(GEA-co-Fru)]. The endocytosis and biological distribution of the functionalized nanoparticles were observed by confocal laser scanning microscopy, and the tumor-killing effect was verified by in vitro cytotoxicity and animal experiments.

【Results】

TEM images (Fig.1) show that GMP and Clof@GMP complex both self-assembled into the nanofibers with length up to micrometers. The addition of P(GEA-co-Fru) to Clof@GMP transformed the fibrous morphologies into nanospheres with a size of ca. 50 nm.

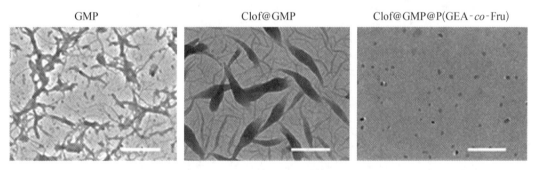

Fig. 1　Representative TEM images (Scale bar: 500 nm)

Excessive fructose can saturate the GLUT5 receptors expressed on the cells through specific recognition and inhibit the binding of P(GEA-co-Fru) to the cell surfaces. In Fig.2, the red fluorescence of Clof@GMP@P(GEA-co-Fru) was significantly weaker in the presence of fructose than in the absence of fructose, which indicated Clof@GMP@P(GEA-co-Fru) was beneficial to be recognized and to be uptaken by the MCF-7 cancer cells.

The whole-body imaging results revealed tumor-targeting capabilities of the Clof@GMP@P(GEA-co-Fru) nanoparticles, promising therapeutic effects of clofarabine to the tumor site.

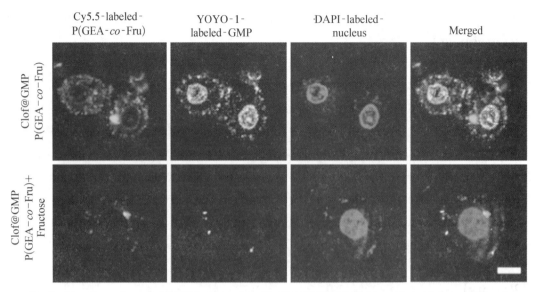

Fig. 2　Fluorescence images of MCF-7 cells treated with Clof@GMP@P (GEA-*co*-Fru) (Scale bar: 10 μm)

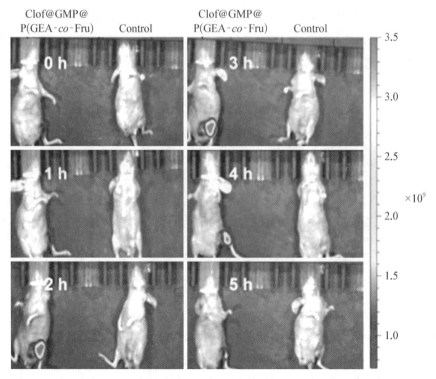

Fig. 3　*In vivo* fluorescence signals obtained from whole-body imaging after intravenous injection

In vitro cytotoxicity experiments indicated an excellent cancer cells-killing effect of Clof@GMP@P (GEA-*co*-Fru). The animal experiments showed that Clof@GMP@P (GEA-*co*-Fru) inhibited the tumor growth most effectively, which is consistent within vitro cytotoxicity tests.

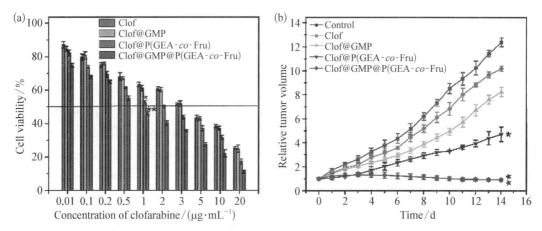

Fig. 4 (a) Cytotoxicity of different samples against the clofarabine concentrations in MCF-7 cancer cells; (b) Relative tumor volume changes in tumor-bearing mice during the treatments

In summary, abundant fructose units in P(GEA-co-Fru) enabled the self-assembled nanoparticles to target MCF-7 cancer cells, which facilitated the accumulation of Clof@GMP@P(GEA-co-Fru) at the tumor site to suppress the tumor growth. This work provides a novel and effective avenue for the design and construction of vehicles for targeted delivery of hydrophobic drugs and tumor therapy.

High Performance PLA Composite with Toughness and Flame Retardancy

Wei Zechang, Fu Yu*

Nanjing Forestry University

*Corresponding author: fuyu@wsu.edu

【Introduction】

Designing polylactic acid (PLA) composites featuring both flame retardancy and excellent mechanical toughness is a great approach to broadening their applications. The halogen-based flame retardants are an efficient choice to improve the flame retardancy of PLA composites. But the pernicious gases and other harmful substances generated from the combustion of halogen-based flame retardants can pollute the environment. To increase its toughness, PLA always blends with bio-based elastomeric polymers and maintains its biodegradability.

【Experimental】

The dual-2D graphene-derived complex (d2D-Gc) of polyester modified graphene (PG) and Layered double hydroxide (LDH) was used to improve the flame retardancy of the PLA composite. Polybutylene succinate (PBS) was used to improve the toughness of PLA composites. Mechanical testing machine and microcalorimeter were used to analyze the effects of the complex of polyester modified graphene (PG) and Layered double hydroxide (LDH) and Polybutylene succinate on PLA composites.

【Results】

As shown in Fig.1, with the addition of PBS, the elongation at break of PLA-PBS/PG-LDH was higher than PLA. As shown in Fig. 1, the PG-LDH had an effect on the flame retardancy of the PLA composite, the heat release rate (HRR) was lower than the PLA. Due to the graphene and LDH were halogen-free flame retardant, the harmful gas was not generated from the combustion of PLA composite. PBS and PG-LDH were also consistent with the idea of sustainable development.

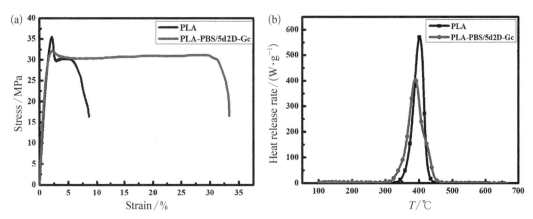

Fig. 1　(a) Tensile properties of PLA composites; (b) HRR curve of PLA composites

Light-driven Liquid Crystalline Networks and Soft Actuators with Degree-of-freedom-controlled Molecular Motors

Lan Ruochen*, Yang Huai

Peking University
*Corresponding author: lanruochen@.pku.edu.cn

【Introduction】

The design and fabrication of photomechanical soft actuators have attracted intense scientific interests because of its potential in the manufacture of untethered intelligent soft robots and advanced functional devices. Liquid crystalline network (LCN) is a superior candidate due to it combines the elasticity of the polymer network and anisotropy of liquid crystal. However, up to date, most of the photo-directed LCN materials are based on the photoisomerization of azobenzene. Exploring and introducing novel photoactive molecules with peculiar advantages into the LCN matrix to fabricate photomechanical LCN actuators with high performances is highly desirable but challenging. Here, the trifunctional and monofunctional polymerizable molecular motors were judiciously designed and synthesized. Novel light-driven liquid crystalline networks (LCN) were prepared by crosslinking overcrowded-alkene-based molecular motors with different degree-of-freedom into the anisotropic LCN. The light-driven molecular-motor-based LCN soft actuators were demonstrated to function as a grasping hand, where the continuous motions of grasping, moving, lifting, and releasing an object were successfully achieved. This work may provide inspirations in preparation of new-generation photoactive advanced functional materials toward their wide applications in areas of photonics, optoelectronics, soft robotics, and beyond.

【Results】

The chemical structures of the molecular motors were shown in Fig.1(a). As demonstrated in Fig.1(b) and 1(c), the photoisomerization and thermal helix

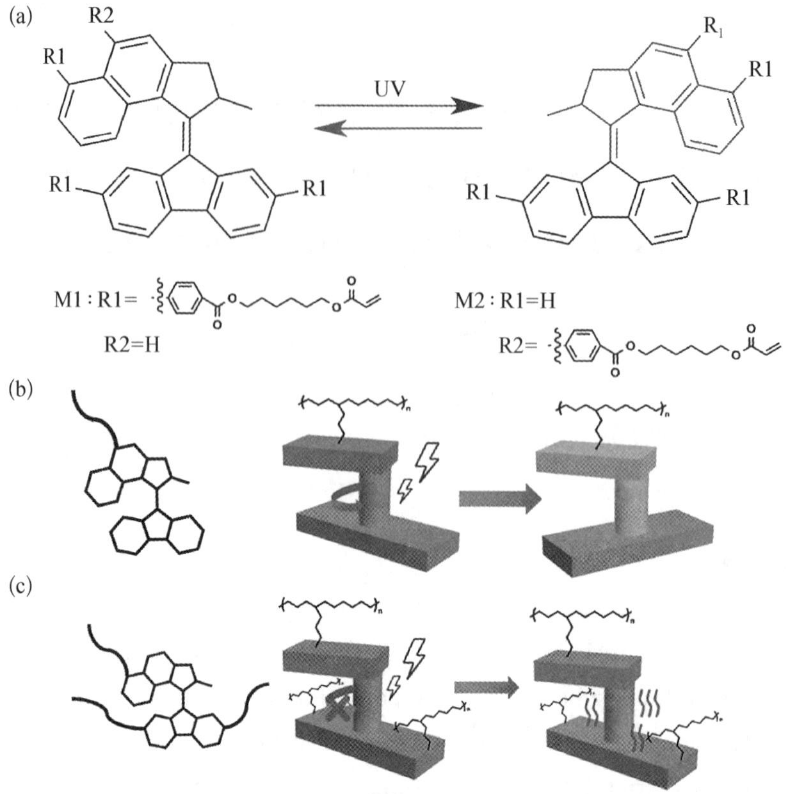

Fig. 1 (a) Chemical structures and schematic illustration of photoisomerization of the molecular motors; (b) (c) Photo-responsivity of the molecular motor with different degree-of-freedom

inversion of the light-driven molecular motor was reversible when only the upper part of the molecular motor were linked to the network, endowing the LCN film with a remarkable photoactive performance. However, photochemical geometric change of light-driven molecular motor fails to work after crosslinking both the upper and lower part of the motor by polymer chains. Interestingly, it was found that the fastened motor could transfer the light energy into localized heat instead of performing photoisomerization. In further investigation, it was found that the heat generated by the fastened molecular motor is high enough to drive the macroscopic motion of the LCN film. As presented in Fig. 2, the LCN film crosslinked by the molecular motor can perform light-directed curling motion due to the striking photothermal effect. Besides, a grasping hand is prepared by molecular motor-based LCN film, the artificial hand can catch the object and then lift, move, and release the object controlled by light.

In conclusion, this work provides a strategy for regulating molecular motion by the macroscopic network, gaining new insight into the interaction between polymer network and molecular-scale motion. It is expected to pave an avenue for the

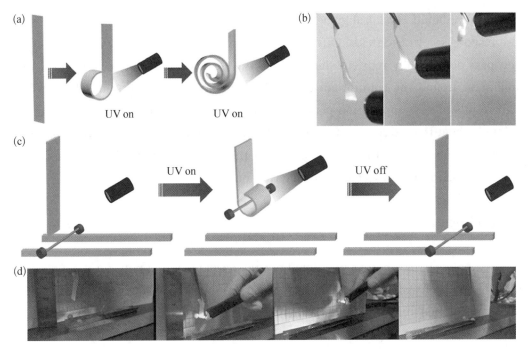

Fig. 2 (a)(b) Schematic illustration and photograph of the photo-responsivity of LCN film with the fastened molecular motor; (c)(d) Schematic illustration and photograph of the photo-driven grasp hand based on the molecular motor LCN film

development of next-generation advanced functional polymeric materials toward their wide applications in areas of photonics, optoelectronics, untethered soft robotics, and beyond.

Responsive Liquid Crystal Network Coatings and Their Applications

Feng Wei*, Liu Danqing, Broer Dirk

Eindhoven University of Technology, Eindhoven, the Netherlands
* Corresponding author: w10.feng@gmail.com

【Introduction】

Liquid crystals are renowned for the application in Liquid Crystal Displays (LCDs). The liquid crystals align with the electric field line and reorient themselves in response to the electric field, constituting the general mechanism of LCDs. With freezing the liquid crystal alignment in the polymer network, liquid crystal networks also respond to electric fields. Different from low-molecular-weight LCs, liquid crystal networks do not fully reorient themselves with electric fields but oscillate in a small amplitude around the equilibrium under the stimulation of alternating electric field. This results in new application perspectives of liquid crystal networks. In this talk, I will talk about the electric field-induced surface topographical deformation of LCNs by elaborating on the actuation mechanism and the resulting applications, for example, remote dry-cleaning, switchable surface frictions.

【Experimental】

We polymerized the reactive liquid crystal acrylates into a polymeric network and investigated its response to electrical signals. The electric field is provided via an array of interdigitated electrodes and the electrically induced surface topographical deformation process is probed with digital holographic microscopy in real-time. This topographical deformation is useful in dust removal from the tilted stage and control of the surface friction of the LCN coating.

【Results】

The LCN coating with a fingerprint configuration is fabricated on a substrate precoated with an array of interdigitated electrodes (Fig. 1a, b). When subjected to an alternating electric field, the polar LC mesogens oscillate with the field, resulting in extra free volume and order parameter decrease. The liquid crystal network would shrink along with the LC director and expand in the perpendicular direction (Fig.1c). The homeotropic domains, therefore, shrink in the z-direction while the resulting lateral stress is absorbed by the planar areas, adding to the expansion of planar areas in the z-direction (Fig.1d). The deformation of the LC network manifests as a topographical deformation of the coating (Fig.1e).[1]

The topographical deformation of the coating surface can also be engineered into dual-responsive to UV light and electric field by adjusting the chemical compositions of LC monomers[2]. The deformation triggered by either stimulus can interact with others. While applied simultaneously, the deformation amplitude caused by either stimulus adds up to the other. When the stimuli are removed from the system, the topographical deformation behavior depends on not only the currently applied stimulus but also on the stimulation history, which is analogous to the sequential logic circuit.

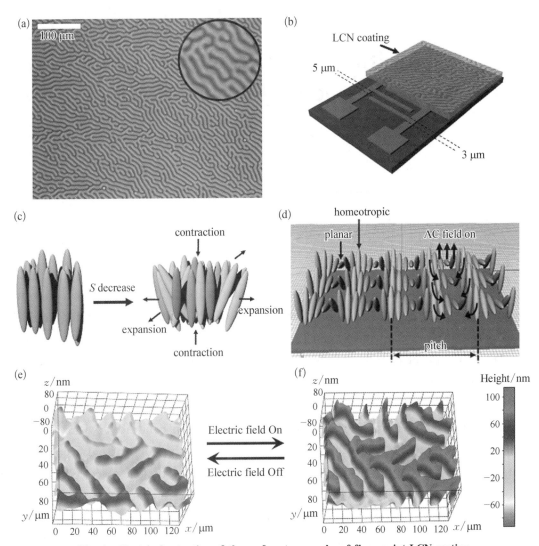

Fig. 1 Electrical actuation of the surface topography of fingerprint LCN coating

References:

[1] Feng, W., Broer, D. J., Liu, D. Oscillating Chiral-Nematic Fingerprints Wipe Away Dust[J]. Adv. Mater., 2018, 30(11): 1704970.

[2] Feng, W., Broer, D. J., Liu, D. Combined Light and Electric Response of Topographic Liquid Crystal Network Surfaces[J]. Adv. Funct. Mater., 2020, 30(2): 1901681.

Controlling the Cell Morphology in Tissue Engineering Bilayer Scaffold

Yu Xi, Zhang Kunxi, Yin Jingbo*

Shanghai University
* Corresponding author: jbyin451@163.com

【Introduction】

The temporomandibular joint (TMJ) supports daily oral activities. The relevant diseases due to trauma or aging remains a clinical challenge despite surgical intervention. Tissue engineering can facilitate the healing of defects of TMJ condyle cartilage through the combination of biomaterials and stem cells. However, few works to date have taken notice of the hierarchical structure of the condyle articular cartilage. Here, a bilayered integrated scaffold is designed to reconstruct the different matrix of TMJ condyle cartilage under the same induction condition, with the final goal of tissue repair. To achieve the bilayered integrated scaffold, poly (L-glutamic acid)-g-polycaprolactone (PLGA-g-PCL) and polyethylene glycol (PEG) was used to construct two layers of the network in a whole by regulating the crosslinking degree of each layer. It was found that the top layer was more hydrophobic than the bottom one after measuring their swelling ratio and distance between hydrophobic domains, which induced that cells would show a flat and spreading status in the top layer and form spheroid in the bottom layer. These two morphologies of cells showed different gene expression levels under the same chondrogenic condition after 14 days. Collagen type Ⅰ (COL Ⅰ) in the top layer was 3.3 folds higher than that in the bottom layer. And collagen type Ⅱ (COL Ⅱ) and Sox 9 in the bottom layer were 1.67 and 3.7 folds higher than in the top layer.

【Experimental】

Preparation and Characterization of the Poly (L-glutamic acid)-graft-Poly (ε-caprolactone)-poly (ethylene glycol) (PLGA-g-PCL-PEG) Hydrogels and Porous Scaffolds

To obtain bilayered integrated tissue engineering porous scaffolds, the initial gel was prepared by crosslinking PLGA-g-PCL copolymer with PEG400 in 2 mL DMSO under EDC chemistry. The bottom layer with 0.027 4 g PLGA-g-PCL copolymer and 0.02 g PEG400 was pouring into the cylindrical mold first. After curing for a few seconds, the top layer with 0.027 4 g PLGA-g-PCL copolymer and 0.015 g PEG400 was pouring onto the bottom layer. Crosslink was continued for 24 h. The gel was then dialyzed against deionized water gradually to turn into a hydrogel which was frozen at $-20\ ^{\circ}\text{C}$ for 8 h and lyophilized for 24 h. Besides, the gels with only one layer were used as control.

Protein Labeling, Incubation, and Retention in Porous Scaffolds and Cell culture *in Vitro*

A solution of $2\ \text{mg} \cdot \text{mL}^{-1}$ of protein (bovine serum albumin, BSA) in 0.1 M sodium carbonate buffer was prepared at first. Then, FITC was dissolved in DMSO at $1\ \text{mg} \cdot \text{mL}^{-1}$ and 0.5 mL FITC solution was added in every 1 mL of protein solution very slowly with gently and continuously stirring. After that, the reaction solution was incubated for 8 h in dark at 4 ℃. NH_4Cl (ammonium chloride) was added to reach a final concentration of 50 mM and the solution was stirred for another 2 h at 4 ℃ to quench the conjugation reaction. The FITC conjugated protein solution was dialyzed against deionized water using a dialysis bag with $3\ 500\ \text{g} \cdot \text{mol}^{-1}$ MWCO for 3 days in the dark. Deionized water was changed every 3 h. At last, $0.2\ \text{mg} \cdot \text{mL}^{-1}$ FITC labeled protein solution was obtained to incubate with top and bottom porous scaffolds (4 mm×4 mm×2 mm) in a 96-well plate for 24 h and 1 mL protein solution each well.

The protein-scaffold system was put in PBS in the dark at room temperature and observed under confocal laser scanning microscopy (CLSM) (FV-1 000, Japan) every 30 min to explain the protein retention in porous scaffolds.

Cells were passaged two to three times before the experiment, and the scaffolds were immersed into gBMSCs (5×10^7 cells $\cdot$ mL^{-1}) with pressure reduction to ensure infiltration and full attachment of cells into the porous structure. The scaffolds used had been sterilized by γ-ray before cell seeding. And the cell-containing scaffolds were then incubated in the culture medium. To observe the morphology of cells, the gBMSCs were pre-labeled with fluorescent 3,3-dioctadecyloxacarbocyanine perchlorate (Dio) dye (Molecular Probes, USA) at 37 ℃ for 20 min before seeding. The labeled cells were then seeded on the scaffolds as described above. The gBMSCs distribution in the scaffolds was observed using confocal laser scanning microscopy (CLSM) (FV-1 000, Japan).

【Results】

In this study, a tissue engineering bilayer scaffold was developed to reconstruct

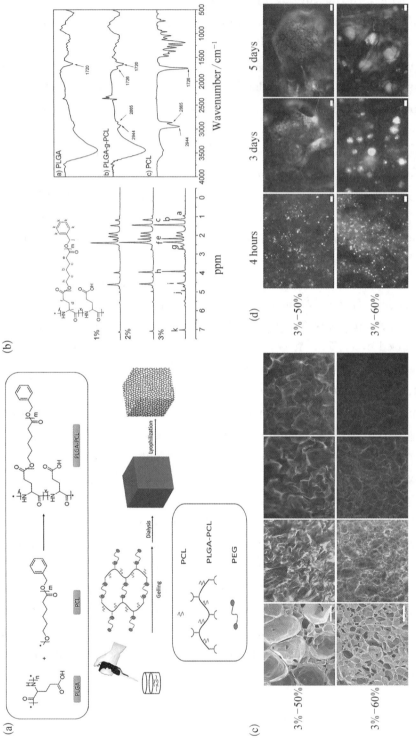

Fig. 1 (a) Schematic of scaffold preparation; (b) ¹H NMR and FTIR of PLGA-g-PCL copolymer; (c) BSA retention of two scaffolds, Bar: 200 μm; (d) In vitro cell culture in two scaffolds; Bar: 50 μm

the hierarchical structure of TMJ condyle articular cartilage. Data showed that the porosity and pore size of two layers of scaffolds were both suitable for nutrient transportation and cell growth. The water affinity of two layers was significantly diverse proved by swelling ratio, the distance between microdomains and ΔH. The top layer was much more hydrophobic while the Bottom one was more hydrophilic, which had a great influence on protein adsorption and, eventually, cell status. The stem cells adhered to the pore walls of the top layer, while in the Bottom layer, the interaction between cells and scaffold was attenuated due to the hydrophilic property. Thus, the stem cells formed cell spheroid finally. Cell status plays an important role in the differentiation of stem cells. Under the same in vitro chondrogenic induction, cells in two layers expressed the feature of fibrocartilage and hyaline cartilage respectively. Based on the above results, the bilayer scaffold was believed an intelligent material in TMJ condyle cartilage tissue engineering, which can adjust the cell status spontaneously and mediate the differentiation of stem cells under the same chondrogenic induction.

第六部分　纤维与低维材料
Section 6　Fibers and Low-dimensional Materials

基于低维材料的人体热管理技术

郭洋[1]，吴波[1]，侯成义[1]，张青红[2]，李耀刚[1]，王宏志[1,2,*]

1 东华大学材料科学与工程学院，纤维材料改性国家重点实验室，上海，201620　2 东华大学先进玻璃制造技术教育部工程研究中心，上海，201620

* 通讯作者：wanghz@dhu.edu.cn.com

人体热管理是指，以服装为载体，通过新的材料与技术对体域热量进行调节和利用。通过人体热管理，一方面，能够使人体保持在健康体温左右，保证人体处在一个安全舒适的状态，并极大程度地提高人体应对外界环境变化的能力；另一方面，能够通过能源转换技术捕获体表多余热量并转换成电能，为随身电子设备提供能源补充。

报告人团队，设计了一种超薄结构的石墨烯纸，并将其应用于人体热管理。研究人员采用高通量的刮涂法和基于维生素 C 的热液还原法制备出了力学性能优异的超薄结构石墨烯纸（1~8 μm）。这种石墨烯纸具有自支撑、可水洗、对皮肤友好的特点。研究人员将这种石墨烯纸应用于服装上，在热的环境下，基于石墨烯材料高热导的特性它能够将服装内环境的热量快速导向外环境，帮助人体体表环境降温；在冷的环境下，基于高电导的特性，它能够在 8 s 内 3.2 V 电压下快速升温至 38℃，帮助人体保暖。为提高穿着舒适性，解决透气性问题，研究人员通过与传统织物结合，利用平纹编织、混编、镂空、剪纸等技术设计了多种热管理器件（图 1a）。这种基于石墨烯纸的热管理器集双向调温功能于一体，实现了降温与保暖一体化设计，能够为未来智能服装的开发提供灵感。

在人体体域热量捕获方面，申请人团队采用锂离子插层法制备了二维 MoS_2 材料，并通过 Au 纳米颗粒的原位生长，在二维 MoS_2 组装过程中引入"肖特基接触"，使得 n 型的 2H 相 MoS_2 发生了 p 型转变，显著改善了混合相中 n 型纳米片对热电性能的削弱，提高了二维 MoS_2 热电材料的功率因子。基于此种材料，报告人团队以纤维织物为基底，构筑了腕带式的演示型热电器件，该器件通过利用人体与外部环境的温差（约 5 ℃）可实现 2.4 mV 的输出电压（图 1b）。为了进一步提升热电转换性能，申请人针对性能优异的传统陶瓷基热电材料，引入了降维的思路，采用溶剂热法制备了 Bi_2Te_3 和 Sb_2Te_3 二维纳米片。在此基础上，通过与石墨烯和碳纳米管复合，构筑了兼具良好柔韧性和优异热电性能的薄膜材料。所设计的柔性热电器件展现出了优异的热电转换性能。

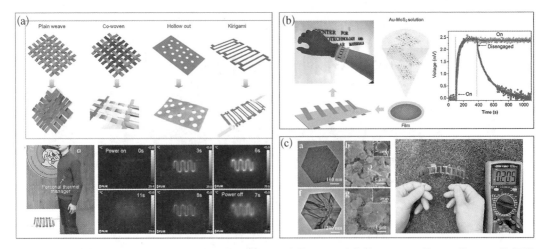

图 1 人体热管理(PTM)：(a) PTM 多设备设计及可穿戴 PMT 设备演示；(b) 基于二维 MoS_2 的柔性 TE 器件的制作及输出电压；(c) Bi_2Te_3、Sb_2Te_3 及其与还原氧化石墨烯(rGO)和碳纳米管(CNTs) 的复合薄膜以及可穿戴 TE 设备的微观形貌

多尺度取向半导体纤维及逻辑器件应用

王 刚 *

东华大学纤维材料改性国家重点实验室，材料科学与工程学院
* 通讯作者：gwf8707@dhu.edu.cn

【引言】

继早期晶体管及超大规模集成电路后，柔性逻辑器件日益引起人们的关注，在航天、军事、医疗及消费电子等领域展现出广阔的应用空间。共轭聚合物材料由于其本征柔性及逻辑响应特性而成为材料的最重要选择。但现有的柔性逻辑器件普遍存在着半导体响应性能较低（载流子迁移率较低）的问题，本文提出了将"多尺度取向半导体纤维"的调控思路应用于柔性逻辑器件的设计。

微观：取向分子链——载流子快速迁移通路；

介观：取向链段聚集体——本征的柔韧性；

宏观：纤维——可融合工业织造特性。

基于以上思路，我们对聚合物分子链高取向结构控制、柔性晶体管印刷设备及方法开发、单纤维半导体 0-1 运算逻辑电路器件等领域进行了系统研究，并探索了其在智能服装、软体机器人等领域的应用（图 1）。

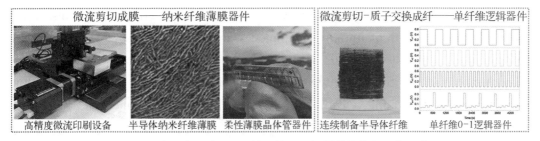

图1 聚合物半导体微纳结构调控及逻辑器件结构与形态设计

【结论】

通过微流效应的引入能够调控聚合物分子链的结晶度和取向性,同时微流的流形设计被证明是一种有效的共轭聚合物微纳结构设计方法;通过剪切印刷的方法能够获得不同共轭聚合物纳米纤维聚集体,为柔性电子的设计提供了多样化的材料选择;搭建了湿法纺丝线,实现了高力学强度、高温度稳定性($-200 \sim 350\ ℃$)、优异0-1逻辑响应功能半导体纤维的连续制备;通过器件的集成化设计,实现了印刷晶体管器件在软体机器人及智能纺织品领域的应用,为新型智能化响应电子器件的设计提供了可能。

Functional Modification and Recycling of Fiber Separator Materials

Xu Guiyin*

Department of Nuclear Science and Engineering, Massachusetts Institute of Technology, Cambridge, MA 02139, USA

* Corresponding author: xuguiyin@mit.edu

【Introduction】

Fiber separator materials have widely practical applications. Here, we designed and prepared several functional separator materials for energy and environment applications. Lithium-sulfur (Li-S) batteries are considered to be one of the most promising energy storage devices in the next-generation high energy power system, owing to their high specific capacity ($1\,675\ mAh \cdot g^{-1}$) and energy density

(2 600 Wh · kg^{-1}).Lithium-sulfur battery is impeded by low utilization of insulating sulfur, dendritic growth of lithium metal anode, and transport of soluble polysulfides. By coating functional materials onto commercial polypropylene separator, we demonstrated the new composite separators that could confine the polysulfides within the cathode side, forming a catholyte chamber, and at the same time block the dendritic lithium in the anode side. Moreover, we also applied our designed separator materials for water treatment and air purification.

The fast development of fiber materials results in a mass of white wastes. Thus, we designed a CMP device, which is tailored towards recycling fiber materials. The CMP process can be used as a feeder for the Fischer-Tropsch process as a method to convert fiber materials into hydrocarbons such as ethanol, diesel, and natural gasses.

Functional Polyelectrolyte Materials Based on Poly(1,2,4-triazolium)s

Zhang Weiyi[1*], Sun Jianke[2], Yuan Jiayin[3]

1 Dong hua University 2 Beijing Institute of Technology
3 Stockholm University, Sweden
* Corresponding author: wyzhang@dhu.edu.cn

【Introduction】

The past decade has witnessed considerable efforts of scientists in addressing challenges in the fields of energy, sustainable development, clean water, and healthy life, and tremendous materials with specific features and advanced applications have been developed. Porous polyelectrolytes, as one of these powerful multifunctional materials platforms, have exhibited the unique strength in solving concerning topics, *e.g.*, sustainability and environmental detection. As one subclass of porous polyelectrolyte, porous poly(ionic liquid)s (PILs) membranes have attracted increasing interest among polymer and materials field owing to their typical applications in sensing, actuation, carbon precursors, *etc*. However, the exploration of the limitation and interesting application scenarios of such porous materials are still appealing.

Recently our group developed a novel poly(ionic liquid)s — poly(1,2,4-triazolium) PILs. After the synthesis of poly(1,2,4-triazolium)s *via* straightforward free radical polymerization, the metal ion loading of such polymers was tested. It is proved that owing to a possible carbene intermediate process, metal clusters down to the size of 1 nm can be stabilized by poly(1,2,4-triazolium)s in organic solutions. Inspired by this carbene chemistry, we designed and developed a 1,2,4-triazolium-based porous polycarbene-bearing membrane. This free-standing nanoporous polycarbene-bearing membrane can serve as an actuator sensor to detect acetic acid (CH_3COOH) down to ~ 3.7 ppm in aqueous media, and could *in-situ* monitor an entire proton-involved chemical reaction event, indicative of their futuristic real-life application in reaction technology.

【Results】

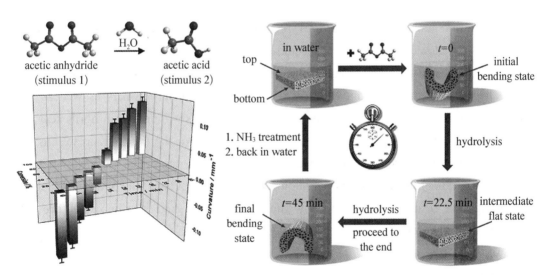

Fig. 1 The cascade actuation of tandem-gradient Ptriaz-TA membrane for monitoring a chemical reaction

Building Gradient Structure in Polymer Blend Fiber and Its Application

Pan Dan[1,*], Chen Long[1], Sun Baozhong[2]

1 State Key Laboratory for Modification of Chemical Fibers and Polymer Materials, College of Materials Science and Engineering, Donghua University, Shanghai 201620 2 College of Textiles, Donghua University, Shanghai 201620

* Corresponding author: dan.pan@dhu.edu.cn

【Introduction】

Bioinspired gradient microstructures provide an attractive template for functional materials with tailored properties. In this study, filaments with gradient microstructures are developed by the melt-spinning of immiscible polymer blends. The distribution of the gradient morphology is shown to be controlled by the viscosity ratio of polymers as well as the geometry of the capillary die. Distinct microstructure gradients with long thin fibrils near the surface region and short large droplets near the center region of the filament, as well as the inverse pattern, are formed in systems with different viscosity ratios. The shear flow field in the capillary can elucidate the formation mechanisms of gradient morphologies during processing. The results demonstrate how the features of a gradient microstructure can be tailored by the design of capillary geometry and processing conditions.

【Results】

For all studied cases of high viscosity ratio PP/PS blend fibers, the average radii of PS domains in the central region are larger than those in the surface region. Higher AR strengthens the gradient structure of PP/PS blend fibers. The morphology of low viscosity ratio PP/PS blend fibers, on the other hand, shows an inverse gradient microstructure with a small diameter of PS phases in the center region and large irregular PS domains near the surface region. Some irregular domains are also observed in the high-viscosity ratio samples, however, they are not as dominant as in the low-viscosity ratio samples. The viscosity ratio can be utilized as an effective parameter to control the gradient morphology in a given capillary die setup. It is interesting to note that the radial distribution of average radii of PS domains displays a W-like profile for the low viscosity ratios. Once the PS domains are shaped under the effect of converging flow in the entrance zone and flow into the capillary, PS domains in different radial positions of the die will experience different shear rates. In general, the low viscous PS phase migrates to the die wall where the shear rate is highest. These PS domains near the surface region will subsequently coalesce into larger domains and form irregular shapes. The PS domains near the capillary center with weak shear rates merely experience weak deformations. In between these two extremes, a competition between breakup and coalescence forms the W-profile in the cross-sections. Gradient PP/PS blend fibers were prepared by melt-spinning with distinct inverse gradient microstructures in bulk. The dispersed phase in the high viscosity ratio system showed long thin fibrils in the surface region and large deformed droplet near the center region. On the contrary, the PS domains in the low viscosity ratio system form irregular domains near the surface region and long thin fibrils near the center region. These two opposite gradient microstructures result from the parabolic shear rate profile in the capillary die due to differences in hydrodynamics of PS domains in low viscosity and high viscosity PP matrix. These features are mainly controlled by the viscosity ratio as well as the die geometry. Thus, the viscosity ratio can act as a powerful tool to manipulate the gradient morphology in a given processing setup.

基于材料表面图案化技术揭示细胞黏附、增殖与分化的影响因素

姚 响[*]

纤维材料改性国家重点实验室,东华大学材料科学与工程学院,上海,201620

* 通讯作者：yaoxiang@dhu.edu.cn

在生物材料学、细胞生物学、组织工程和再生医学中,研究并揭示细胞—材料相互作用中各种因素的作用效应和机理都极为重要,也是生物医用材料领域的一个重大共性科学问题。在传统的细胞培养体系中,影响细胞-材料相互作用的多种因素(细胞因素、细胞外基质因素和可溶性因子)往往混杂在一起,难以区分。得益于先

进材料的发展,材料表面图案化技术可用于构建具有细胞黏附反差特性的图案,进而达到对细胞黏附的精确调控。该技术有望用于单独研究不同因素对细胞黏附、增殖和分化等行为的影响,从而为深入准确地理解细胞-材料相互作用提供独特的方法[1]。

利用先进的聚乙二醇水凝胶图案化技术,我们实现了对"单个"细胞黏附铺展持续21天的持久控制。通过铺展面积对细胞黏附行为的详细考察,我们提出并证实了细胞黏附相关特征面积的存在(例如细胞由凋亡到存活转变的特征性铺展面积 A^*,由单细胞黏附到多细胞黏附转变的特征性面积 A_{C2}),发现不同细胞具有相似的黏附规律[2];通过铺展面积对单细胞增殖行为的详细考察,我们提出并证实了细胞增殖相关特征面积的存在(例如细胞由几乎不能增殖到能发生受限增殖转变的特征性铺展面积 A_P,细胞由受限增殖到几乎能发生自由增殖转变的特征性铺展面积 A_{FP}),并证实铺展面积对细胞增殖行为的影响具有明显的细胞种类依赖性[2];通过适当微米图案面积和形状的设计,我们获得了大概率的单细胞黏附并实现了对细胞形状的持久控制,进而考察了细胞形状对干细胞成骨和成脂分化行为的影响,并证实形状效应是一个影响细胞分化的固有因素(可独立于可溶性诱导因子发挥作用)[3-5];基于分子手性表面和微米图案的同时构建,我们还仔细考察了L型和D型半胱氨酸表面(分子旋光性表面)对细胞分化行为的影响,并证实材料表面的分子旋光性特征首先导致了蛋白吸附和细胞黏附面积的差异,进而影响了干细胞的成骨和成脂分化[6]。

参考文献:

[1] Yao X, Peng R, Ding J. Cell-Material Interactions Revealed via Material Techniques of Surface Patterning[J]. Advanced Materials, 2013, 25(37): 5257-5286.

[2] Yao X, Liu R, Liang X, et al. Critical Areas of Proliferation of Single Cells on Micropatterned Surfaces and Corresponding Cell Type Dependence[J]. ACS Applied Materials & Interfaces, 2019, 11: 15366-15380.

[3] Yao X, Peng R, Ding J. Effects of Aspect Ratios of Stem Cells on Lineage Commitments with and without Induction Media[J]. Biomaterials, 2013, 34(4): 930-939.

[4] Peng R, Yao X, Ding J. Effect of Cell Anisotropy on Differentiation of Stem Cells on Micropatterned Surfaces through the Controlled Single Cell Adhesion[J]. Biomaterials, 2011, 32(32): 8048-8057.

[5] Peng R, Yao X, Cao B, et al. The Effect of Culture Conditions on the Adipogenic and Osteogenic Inductions of Mesenchymal Stem Cells on Micropatterned Surfaces[J]. Biomaterials, 2012, 33(26): 6008-6019.

[6] Yao X, Hu Y, Cao B, et al. Effects of Surface Molecular Chirality on Adhesion and Differentiation of Stem Cells[J]. Biomaterials, 2013, 34(36): 9001-9009.

可降解的金属有机纳米诊疗试剂的开发和生物应用

余 诺*

纤维材料改性国家重点实验室,东华大学材料科学与工程学院,上海,201620
* 通讯作者:yunuo@dhu.edu.cn

癌症严重威胁人类的健康,2018年全球新发癌症病例1 810万例,死亡人数达960万人[1]。癌症的现有治疗方法主要有手术切除、放射治疗、化学药物治疗、中医药治疗等,虽有一定效果,但其副作用较强。因此开发新型、无副作用的癌症治疗技术受到了国际各国科学家的广泛关注。声动力治疗技术以组织穿透能力强的超声波作为能量,具有特异性和选择性杀伤肿瘤组织的特点,这一无创治疗手段引起了研究者的广泛关注。该方法的基本原理是将声敏剂富集到肿瘤中,利用超声波激发肿瘤,其中的声敏剂吸收能量产生活性氧物质(例如单线态氧),活性氧物质可损伤癌细胞内各种细胞器和DNA,进而诱导癌细胞凋亡[2]。在声动力治疗肿瘤技术中,开发性能优异的声敏剂是基础。

声敏剂目前主要有两种发展方向:半导体和有机小分子。但是,半导体材料具有较低的单线态氧产率,有机分子亲水性极差,导致其只能负载在纳米载体中。为了解决上述问题,我们首次将临床试剂血卟啉单甲醚分子(HMME,来源于血红细胞)和Fe^{3+}离子,自组装形成了金属-有机配位纳米颗粒(图1),它可同时用作新一类声敏剂、药物载体和成像试剂。具有以下创新点:(1) 首次构筑了Fe^{3+}-HMME纳米颗粒作为声敏剂,HMME和Fe都具有良好的生物兼容性;(2) 它展现了优异的单线态氧产生能力和磁共振成像性能;(3) 大的比表面积和多孔结构便于高效负载药物DOX,同时还展现了pH响应的药物释放行为;(4) 声动力和药物联合治疗能明显抑制2 cm深处肿瘤的生长。该新型声动力诊疗平台实现了安全、高效的肿瘤诊疗,并将引导其他声敏剂的进一步开发。

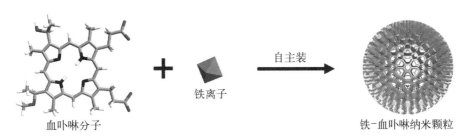

图1 铁离子和血卟啉分子自组装成金属-有机配位纳米颗粒的示意图

参考文献：

[1] Bray F, et al. CA Cancer. J. Clin., 2018, 68: 394.
[2] Chen H, Zhou X, Gao Y, et al. Drug Discov. Today, 2014, 19: 502-509.

Heterogeneous Scorpionate Site in MOF: Small Molecule Binding and Activation

Wang Le

State Key Laboratory for Modification of Chemical Fibers and Polymer Materials, College of Material Science and Engineering, Donghua University, Shanghai, 201620, China

Scorpionate ligands are widely used in coordination chemistry and catalysis. These ligands are usually neutral or mono-anionic, hence the scorpionate transition metal complexes are often bonded to counter anions that blocking catalytic sites. The metal-organic framework reported here, CPF-5, presents a unique di-anionic scorpionate backbone, therefore allows multi-guest molecule binding.

We have demonstrated the CPF-5 scorpionate site can be utilized as a multi-guest binding site for various Lewis bases. CPF-5 could act as a crystalline sponge for single-

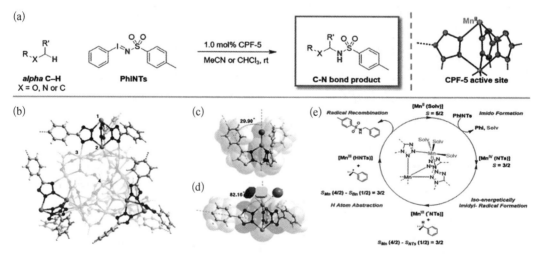

Fig. 1 Catalytic C—H Amination by CPF-5. (a) Amination reaction; (b) Active site structure; (c) Homogeneous scorpionate analogue; (d) Scorpionate site in CPF-5; (e) Proposed reaction mechanism

crystal X-ray structural characterization of a variety of compounds using coordinative alignment, second coordination sphere interactions, or both.

Moreover, the C—H bonds in these binding small molecules could be aminated by the CPF‑5 scorpionate catalytic sites in the presence of nitrene precursors. This Mn-based outer-sphere catalyst promotes the direct amination of C—H bonds in an intermolecular fashion with unrivaled activity producing >105 turnovers. And CPF‑5 could be even recycled and reused, it is found to be the only living catalyst for C—H activation to date.

Bio-Inspired Nanochannel Materials

Meng Zheyi*, Zhu Meifang

State Key Laboratory for Modification of Chemical Fibers and
Polymer Materials, College of Material Science and
Engineering, Donghua University, Shanghai 201620, China
*Corresponding author: mengzheyi@dhu.edu.cn

Based on track-etched PET porous membrane, the preparation, characterizations, and functionalizations of conical/columnar nanochannels are discussed here. By combining PET nanochannels with responsive and energy conversion materials, functions of nanochannels in the energy conversion, and discussed the relationship between ion transport properties of nanochannels and energy outputs of the system have been investigated. The mechanism of regulations in energy conversion with nanochannels will be beneficial for the further application of PET membranes in the energy field.

Firstly, calcein-modified multiporous PET membranes with conical channels in nanofluidic devices are designed to enhance calcium-responsive intensity and stability of ionic currents. Deprotonation of carboxyls on nanochannel surface and binding of Ca^{2+} will change surface charge density, which influences the ionic conductance of nanochannels. So, calcium-binding could exert nano-gating for ionic fluid through the channel. Calcein, a tetradentate ligand with four carboxyl groups, is immobilized in the channel to enhance the response to Ca^{2+}, by binding the calcium ion to form a chelate. And the capability of these carboxyl groups for Ca^{2+}‑binding could be regulated by the deprotonation of the three carboxyls.

Then, inspired by the function of natural ion channels on the photosynthesis membrane, PET membrane with conical nanochannel arrays are introduced into a photoelectrical conversion system to regulate the light-driven ionic current. In this system, photosystem Ⅱ (PSⅡ) protein was applied as the "pump" to convert light into ionic currents, and the PET membrane with conical nanochannel arrays was chosen as the "valve" to regulate the ionic current produced in the system. With the charged inner surface and the asymmetry in shape, PET conical nanochannels could selectively handle molecular or ion species in the fluid.

Besides the two bio-inspired examples above, the structure of gap junction channels also sparks a new design for constructing artificial bi-channel nanofluidic devices and study their pH-dependent ion transport properties. These bi-channel devices were built by combining two PET membranes with columnar nanochannel arrays varying in size or surface charge. The ion current rectification of the bi-channel can be attributed completely to the cooperation of the two participating nanochannels, which simplifies the analysis of the origin of ionic selectivity. Besides PET films with columnar nanochannel arrays are easy to prepare and functionalize, and varying ion-transport properties of bi-channels can be achieved by different arrangements of these PET nanochannels.

On-surface Synthesis and Atomically Precise Fabrication of Low-dimensional Carbon Nanostructures

Sun Qiang*, Oliver Gröning, Pascal Ruffieux, Roman Fasel

Empa, Swiss Federal Laboratories for Materials Science and Technology, nanotech@surfaces Laboratory, 8600 Dübendorf, Switzerland

* Corresponding author: qiang.sun@empa.ch

【Introduction】

Over the last decade, on-surface synthesis has emerged as a powerful tool for the fabrication of novel nanomaterials. The employed ultra-high vacuum (UHV) conditions and the atomically clean, catalytic surfaces used, provide the ideal playground for both synthesis and characterization of atomically precise structures that are hardly accessible through traditional solution-based methods. A significant number of structures/materials which are challenging by conventional solution chemistry could be achieved.

【Experimental】

On-surface synthesis is carried out in an ultra-high vacuum (UHV) chamber coupled to a scanning probe microscopy operated at low temperatures. The structural and electronic properties of the nanostructures are characterized by UHV-based scanning probe methods (STM/STS/nc-AFM) at single-molecule level, complemented by density functional theory and tight-binding calculations.

【Results】

In this work, I will first briefly discuss the emerging on-surface synthesis. Then, several novel low-dimensional carbon nanostructures achieved by our group in collaboration with international research groups will be introduced (Fig.1). Among others, graphene nanoribbons (GNRs), stripes of graphene, are of central interest due to their widely adjustable electronic properties promising for future electronics. Moreover, the formed GNRs can be processed under ambient conditions and incorporated as the active material in a field-effect transistor.

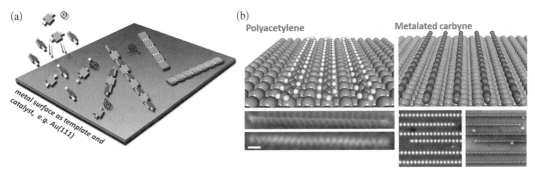

Fig. 1 (a) Schematic of the different steps in on-surface synthesis; (b) Structural models and high-resolution STM/nc-AFM images of individual polyacetylene and metallated carbyne chains

Preparation and Properties of Vanadium Oxides-based Electrodes for Supercapacitors

Sun Wei*, You Zhengwei*

College of Materials Science and Engineering, Donghua University
* Corresponding authors: weisun@dhu.edu.cn; zyou@dhu.edu.cn

【Abstract】

Supercapacitors have attracted great attention in energy storage applications, because of their high specific capacitance, rapid recharging, and the potential of high energy density. Vanadium oxides-based materials are representative of the promising electrode materials and have been widely used[1]. However, the fatal drawbacks of them that confine the further application of their high theoretical capacitance are the dense structures and sluggish ion/electron diffusion. The key to solving these problems is to improve the conductivity of them by integrating conductive carbon materials and constructing rational architectures. Herein, we provide a new strategy to use the defects to direct the vanadium oxides in-situ growth on N-doping carbon scaffolds into three-dimension (3D) composite aerogel based on lattice matching between defect-rich carbon and vanadium oxides. The vanadium oxides nanowires are aligned vertically on porous conductive carbon backbone or uniformly coated on the surface of carbon nanofibers to form 3D porous networks with core/shell nanostructures[2,3]. The first-principle calculations have been performed to reveal the effect of defects on the local electronic structure and G_{ef} of carbon materials, and screen the architecture of carbon with numerous defects, which can also match with the vanadium oxide lattice. We find that the introduction of N atoms into carbon lattice plays an important role in the formation of vertical VO_x nanowires arrays or core/shell structures on carbon materials, which is attributed to different VO_x nucleation processes. First-principles simulations and the relative experiments are managed to demonstrate the effect of incorporating N atoms into conductive carbon backbone on the formation of different vanadium oxides nanostructures on it[3,4]. Such a design offers distinct advantages for vanadium oxides-based materials for supercapacitors. As a consequence, it delivers an ultrahigh specific capacitance of 710 F·g^{-1} (at a current density of 0.5 A·g^{-1}) and exhibits outstanding rate performance and good cycling behavior (after 20 000 cycles 95% retention of specific capacitance)[3]. This study will open a way to guide the fabrication of nanocomposites by combining different dimensional nanomaterials.

References:

[1] J. Zhu, L. Cao, Y. Wu, Y, Gong, Z. Liu, H.E. Hoster, Y. Zhang, S, Zhang, S. Yang, Q. Yan, P.M. Ajayan and R. Vajtai, Building 3D structures of vanadium pentoxide nanosheets and application as electrodes in supercapacitors[J]. Nano Letters, 2013, 13: 5408-5413.

[2] W. Sun, G. Gao, K. Zhang, Y. Liu and G. Wu, Self-assembled 3D N-CNFs/V_2O_5 aerogels with core/shell nanostructures through vacancies control and seeds growth as an outstanding supercapacitor electrode material[J]. Carbon, 2018, 132: 667-677.

[3] W. Sun, G. Gao, Y. Du, K. Zhang and G. Wu, A facile strategy for fabricating hierarchical nanocomposites of V_2O_5 nanowire arrays on a three-dimensional N-doped graphene aerogel with a synergistic effect for supercapacitors[J]. Journal of Materials Chemistry A, 2018, 6: 9938-9947.

[4] W. Sun, Y. Du, G, Wu, G. Gao, H. Zhu, J. Shen, K. Zhang and G. Cao, Constructing metallic zinc-cobalt sulfide hierarchical core-shell nanosheet arrays derived from 2D metal-organic-frameworks for flexible asymmetric supercapacitors with ultrahigh specific capacitance and performance[J]. Journal of Materials Chemistry A, 2019, 7: 7138-7150.

面向高比能锂离子电池的设计策略与关键材料技术

吴英强[1],邹业国[1,2],申亚斌[1,2],王立民[1],明军[1,*]

1 中国科学院长春应用化学研究所   2 中国科学技术大学
* 通讯作者:jun.ming@ciac.ac.cn

【引言】

发展新能源汽车是我国汽车工业智能化的国家战略,其核心技术是发展先进的锂离子电池技术,即如何使能量密度、循环寿命、安全性和成本达到平衡。在锂离子电池的关键部件中,正极对上述四个方面的影响最大。镍钴锰三元氧化物($LiNi_{1-x-y}Co_xMn_yO_2$)因具有理论比容量高(278 mAh·g^{-1})、放电电压平台高、工作温度范围宽等优点而受到广泛关注。尽管三元氧化物正极已实现产业化,但目前更大规模应用面临的瓶颈是如何突破其容量输出与循环寿命、热稳定性和成本之间存在的矛盾。科研人员主要从三个方面进行技术和基础科学研究:一是增加镍含量和降低钴的含量,来获得高的比容量输出并降低成本;二是通过表面包覆和电解液改性,抑制富镍三元氧化物深度脱锂时的表面结构相变和电解液的氧化分解;三是融合先进表征技术和计算模拟等手段研究晶体结构演变、电极/电解液界面的热力学/动力学特性。本文首先综合考虑了锂离子电池正负极的容量发挥、容量匹配、首次库伦效率、结构设计、电极工艺与电解液用量等关键参数,经数学推导建立一个新的估算电池能量密度的经验模型(图1),然后据此开展针对性的关键材料研究。

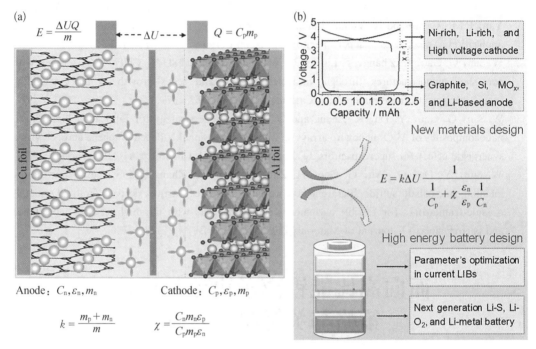

图 1　锂离子电池能量密度估算经验模型及其应用

【实验方法】

采用共沉淀-控制结晶合成高密度的镍钴锰三元氧化物正极材料,并进行表面包覆磷酸铝,提高界面晶体结构稳定性及电化学性能。通过粉末 X 射线衍射(XRD)、扫描电镜(SEM)、投射电镜(TEM)、X 射线光电子能谱(XPS)和电化学交流阻抗谱(EIS)分析镍钴锰三元氧化物在循环过程中界面晶体结构演变对电化学性能的影响。

【结果】

根据能量密度的计算公式,经过数学推导得到一个新的评估电池能量密度的经验模型。利用该模型能快速评价和筛选适用于不同能量密度电池设计的正负极材料特性,并提出可实现>300 Wh·kg^{-1}电池的方案(图1),其中正极材料的物理化学性质(密度、比表、表面结构、容量发挥和热稳定性等)对锂离子电池的能量密度和循环寿命影响最为关键。据此,通过共沉淀-控制结晶合成高密度的镍钴锰三元氧化物正极材料,并发展了新型的非水相薄膜磷酸铝包覆方法实现 NCM333 的超薄磷酸铝的表面包覆(图 2),进一步结合 STEM 和 EELS 研究包覆后界面晶体结构和元素价态的演变,以及其与电化学性能的构效关系。

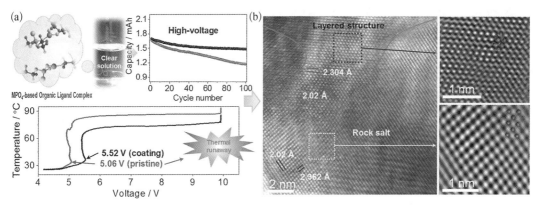

图 2　高压三元材料的表面包覆、电化学性能及循环过程中的结构退化分析

第七部分 新材料技术产业链

Section 7　Industrial Chain for New Materials

Plain Low Carbon Steel Resisted to Corrosion via RE Alloying

Lian Xintong[1], Liu Zheng[1], Fan Jianwen[2], Liu Tengshi[1],
Xu Jing[1], Xu Dexiang[1], Li Jun[1], Dong Han[1,2,*]

1 Shanghai University   2 Central Iron & Steel Research Institute
* Corresponding author: 13910077790@163.com

【Introduction】

Corrosion is one of the main failure modes of plain low steels. One have been trying to find economical and feasible solutions to solve this problem. La, Ce, and Y are the surplus elements in the mining and use of RE elements in China. The development of corrosion resistance steel alloyed with RE elements can not only solve the redundancy in the mining and use of RE but also improve the duration of steels. Researches have shown that adding RE elements into steels can effectively improve corrosion resistance. However, two important problems need to be solved. One is to ensure the effective addition method and stable high yield of RE elements in current continuous steel processing, and another is to re-recognize the prevailing theories of rare earth alloying.

【Experimental】

Industrial hot-rolled coils of plain low carbon steels (70% of total steel output) and low alloy steels (20% of total steel output) were produced by effective RE alloying technologies. Electrochemical tests, salt spray corrosion tests, and wet-dry cycle immersion tests were conducted to compare the change of corrosion resistance. AES, EPMA, SEM, and XRD were used to analyze the effects of RE on corrosion morphology and rust layer structure.

【Results】

Due to atom radius difference between Ce and Fe, Ce tend to segregate towards grain boundaries and be enriched at the interface between rust layers and matrix, which promotes the redistribution of P and Cr (Fig. 1). RE also enhances the compactness of rust layers and

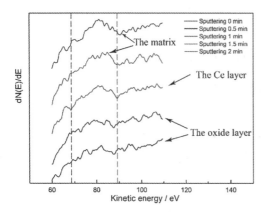

Fig. 1 AES peaks of Ce in 09CuPCrNi steel

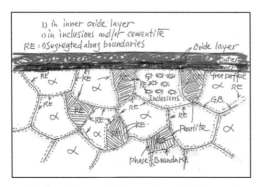

Fig. 2 Schematic illustration of RE in steels

hinders the corrosion ions from contacting the matrix, thus improving the corrosion resistance of steels effectively.

It is believed that RE can modify inclusions and reduce the electrode potential difference between inclusions and matrix, thus avoiding pitting corrosion. RE can also segregate towards interfaces to reduce the interface energy and avoid local corrosion. Furthermore, RE can stabilize and condense the rust layer structure and effectively slow down the corrosion rate (Fig. 2). Whether RE atoms can be dissolved in ferrite or form clusters is a problem to be further explored. An effective method of RE alloying technology to obtain 100 - 300 ppm RE content in steels could guarantee corrosion resistance. A catalogue of novel low carbon steels via RE alloying has been created to resist corrosion and may lead to a scenario of the long duration of steel structures at a lower reasonable cost (Fig. 3).

Fig. 3 Exposed Q235B, Q235RE, Q355B, and Q355BRE samples and Q355BRE used for building

新型低成本耐蚀钢的应用领域及前景

王睿谦[1]，刘超[2]，朱嘉楠[1]，徐德祥[1,3]，廉心桐[1,*]，董瀚[1]

1 上海大学 2 沈阳嘉美科技有限公司 3 上大新材料(泰州)研究院有限公司
* 通讯作者：xtlian@shu.edu.cn

【引言】

低成本、高性能、长寿命是经济型耐蚀钢的主要特点，目前已在上海、辽宁、山东、湖南、江苏等地区开展了新型低成本耐蚀钢的示范工程应用。随着市场共建的逐步扩大并被工程领域不断认可，新型低成本耐蚀钢可形成政府推广—企业制造—百姓自发购买的良性循环。

【内容】

新型低成本耐蚀钢是基于量大、面广的普碳钢和低合金结构钢，通过新型合金化技术显著提高钢材耐腐蚀性能，延长服役寿命并有效控制钢材成本研制的新型钢。团体标准《稀土耐候结构钢》(T/CSM 12—2020)的制定也满足了低成本耐蚀钢对材料标准的需求。

新型耐蚀钢的性能突出，在钢结构建筑领域具有建造时间快、服役寿命长等优点。湖南亿迈文化传媒有限公司就以 S20 耐候钢中板与薄板为主，加工成梁、柱、楼板、屋面板等作为政府形象工程推广，市场潜力巨大。耐蚀钢在地螺丝领域也得到了广泛应用。威海立达尔机械股份有限公司生产的耐腐蚀地螺丝微型钢管桩单面腐蚀率 0.1 mm/年，裸钢性能十分优越。在浅海腐蚀试验中发现，与市场同类产品相比，新型耐蚀钢的使用寿命提升 1 倍多，大幅度降低了成本投入。耐蚀钢在装饰装修行业更是前景巨大，沈阳嘉美科技率先采用经济型稀土耐蚀钢替代传统 Q235 钢材应用于外挂装饰框架，服役寿命显著提高。

新型低成本稀土耐蚀钢解决了钢的防腐问题，是钢结构颠覆性的变革，必将推动钢铁行业实现质的飞跃。同时还应开发配套的连接材料，如耐蚀螺栓，紧固件(6.8～10.9 级)以及适用于各类焊接方式如埋弧焊、气保焊等的配套焊接材料。当前推进新型耐蚀钢的应用势在必行，关于耐蚀钢材料的标准已经制定，亟须推动应用标准的制定，尽早纳入建筑材料规范。

图 1 新型耐蚀钢应用在钢结构装配式住房、耐腐蚀地螺丝、外挂装饰装修幕墙等领域

中国金属学会团标《稀土耐候结构钢》要点解析

廉心桐[1,*],范建文[2],韦习成[1],刘腾轼[1],李钧[1],徐德祥[3],董瀚[1]

1 上海大学 2 钢铁研究总院 3 上大新材料(泰州)研究院有限公司

* 通讯作者:xtlian@shu.edu.cn

【引言】

我国是稀土大国,稀土资源十分丰富。钢中加入轻稀土成本相对较低,但可以促进材料表面形成致密的保护膜,提高材料的耐大气腐蚀性能。在相同耐蚀性条件下,可减少 Cu、Cr、Ni 的加入量,节约贵重合金元素,降低合金成本。据此提出了稀土耐候结构钢标准。此外,可以有效促使夹杂物球化而改善轧材的各向异性,特别是板材的横向性能,可以细化钢的铸态组织。发展稀土耐候结构钢既能够解决钢材严重的腐蚀问题,又能够消耗我国的富裕轻稀土,促进钢材的更新换代。本团标是稀土耐候结构钢产业技术创新基础设施的重要内容,为此由上海大学牵头联合 10 多家单位制定了《稀土耐候结构钢》(T/CSM 12—2020)团体标准,参与单位包括研发生产耐候钢的各企业、高校及科研院所,具有广泛代表性。

【原则及主要技术】

该标准规定了稀土耐候结构钢的尺寸、外形、重量及允许偏差、技术要求、试验方法、检验规则、包装、标志及质量证明书。本标准适用于车辆、桥梁、集装箱、建筑、塔桅和其他结构用具有耐大气腐蚀性能的热轧钢板和钢带、型钢和长材。稀土耐候钢可制作螺栓连接、铆接和焊接的结构件。部分技术要点如表1所示。

表1 本标准的主要技术要点

章条编号	技 术 内 容	要 点 解 析
3	**稀土耐候结构钢Ⅰ类** 在 GB/T700 中的 Q235 普碳钢或 GB/T1591 中的 Q355 低合金高强度钢的基础上,通过添加少量的稀土元素,在金属基体表面形成致密稳定的氧化保护层,耐大气腐蚀性能大幅度提高。 **稀土耐候结构钢Ⅱ类** 在传统耐候钢基础上,通过添加少量的稀土元素,可适当减少合金元素 Cu、Cr、Ni 等的含量,在金属基体表面形成致密稳定的氧化保护层,是满足耐大气腐蚀性能要求的钢。	稀土耐候结构钢的分类根据钢材对不同耐候性能的要求,分为稀土耐候结构钢Ⅰ类、Ⅱ类,钢材成分有所差异。
7.1	RE 表示添加以 La、Ce、Y 为主的一种或多种稀土元素。	加入的 RE 含量可以为单一或混合稀土,总含量不可超过 450 ppm。
8.2	参照 GB/T4336 检验钢中稀土元素含量,需要制作标样。标样可以先采用 GB/T24520 和附录 C 标准对钢样的稀土元素进行检验并以此作为参照 GB/T4336 标准进行稀土元素检验的标样。	稀土含量的分析需要特殊的标样进行检验,保证稀土含量测试的准确性。
附录 B	B.1 钢的耐腐蚀性能根据供需双方协议确定下列两种方法进行评价。	由于耐腐蚀性能测试的周期性及复杂性,一般通过双方协议进行,不作为交货要求。

发展新型稀土耐蚀钢的产业链构建

徐德祥[1,2],王睿谦[2],张新威[3],赵希誉[4],
欧阳烈松[5],朱嘉楠[2],廉心桐[2],董瀚[2,*]

1 上大新材料(泰州)研究院有限公司　2 上海大学　3 威海立达尔机械有限公司　4 湖南中城住工科技有限公司　5 湖南亿迈节能科技有限公司
* 通讯作者：13910077790@163.com

【引言】

2019 年我国粗钢产量近 10 亿吨,超过全球粗钢总产量的 1/2。低成本钢铁材料耐腐蚀技术，特别是针对量大、面广的普碳钢（占比约70%）和低合金钢（占比约20%），将产生巨大的经济和社会效益。"新型基础设施建设"（简称"新基建"）概念自提出以来，耐腐蚀新材料的开发就迫在眉睫，但与新型稀土耐蚀钢有关的问题，包括稀土合金原料、钢材生产流程、产品销售及应用还未作为建筑规范广泛使用，亟须完善新型稀土耐蚀钢的全产业链问题并进行推广应用。

【内容】

钢铁产业作为"新基建"的支柱产业，稀土作为我国的战略资源，采用稀土合金化技术生产的耐蚀钢是基于"新基建"概念落地的新钢铁材料，能够显著提高钢材的耐蚀性能，延长服役寿命，减少后期维护，目前技术成熟度可达7~8级。

新型稀土耐蚀钢已作为政府形象建设的标杆工程落实使用，但其产业链结构有待进一步完善，主要问题是缺少中间环节的经销商。新型稀土耐蚀钢尚未广泛地用于钢材制造，经销商手中稀土钢材产品几乎没有库存，但可能存在短期内下游用户激增的现象。而上游的生产厂家由于排产难度大、批量小规格多等各种原因，导致无法满足下游用户的需求。因此，建立经销商至关重要，经销商可通过各种渠道资源进行分销配送，实现产品的快速销售和市场覆盖。

除此之外，还需进一步完善稀土耐蚀钢产业链条，配套所有产品型号，包括稀土耐蚀钢各类板、H型钢（工字钢）、槽钢、角钢、管材等，形成研发—生产—装配的完整产业链。扩大市场应用范围，增加钢铁生产企业供给数量，降低产品成本，促进中下游企业合作共赢。

从目前来看，新型稀土耐蚀钢的共性问题包括腐蚀机理、服役行为及评价方法研究，亟须形成具有我国自主知识产权的耐蚀钢材料体系。主要包括以实验室为主的加速腐蚀试验、不同地区的户外暴晒试验、应用环境腐蚀试验，通过积累腐蚀数据并进行分析，最终形成完整的腐蚀数据库，为推广新型稀土耐蚀钢奠定基础。

德国螺纹接头正向开发对我们的启示

郑 军[*]

上海兹懋仪器科技有限公司

[*] 通讯作者：javen.zheng@zmart-china.com

【引言】

螺纹接头涉及紧固件、被连接件，也涉及装配、服役，更需要考虑服役的因素、失效机理，还需要依据从实践中获得的经验如安全裕度等,所以是很多方面知识的汇集。面对新材料应用、减重需求、安全舒适需求等新趋势,以往对标以及粗放型的螺纹接头设计理念面临着重大挑战。如何保证可靠的设计、可靠的运行服役是相关从业者的责任和目标。

【我国螺纹接头设计现状】

大多数螺纹接头的设计采用对标标杆车型来选用紧固件,其后设定拧紧扭矩,最终进行耐久验证,验证过程中出现松动异响等现象后,加大扭矩或更改螺栓规格甚至大动干戈去整改被连接件的结构和尺寸。

【结论】

科学的螺纹接头设计模式称为正向设计,是通过试验测试获得实际摩擦系数,测量预紧力损失等信息,通过非线性的计算校核软件——《螺栓设计师专业版》按照初选螺栓、确定拧紧系数;计算最小夹紧力;分解工作载荷/载荷因数;确定最小、最大装配预紧力……直至计算拧紧扭矩等步骤进行计算校核。一般来说,简单的螺纹接头采用这种方法即可保证一定的可靠性设计。针对复杂受力、存在受载变形行为、预紧力损失以及潜在自松行为的螺纹接头,还需利用有限元仿真来进行模拟,以获得进一步的信息,如具体受力、预紧力损失、松动等数据,再输入到计算校核软件中进行计算校核。同时,基于行业设计实践所获得的针对安全因子的选择,也是获得可靠性设计的重要一环。因此,螺纹接头正向设计应整合试验、计算与仿真分析,以此获得可靠的全寿命周期的设计方案。具体流程图如图1所示。

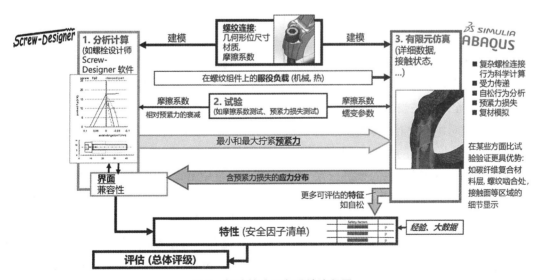

图1　螺纹接头正向设计流程图

汽车紧固件产业链的思考

靳宝宏*,王立新,袁峰,王晓勤,钱鹏,曹培元

泛亚汽车技术中心有限公司
*通讯作者：baohong_jin@patac.com

【引言】

作为"工业之米"，紧固件在各行各业中有着广泛的应用，紧固件产业的发展对我国工业发展非常重要。汽车行业是一个国家高端制造业的标杆，而汽车紧固件产业带动了中国紧固件产业的快速发展。当前，网联化和数字化促进了紧固件产业的升级，贸易争端加速了产业链的本地化进程，紧固系统设计的日趋成熟驱动了坚固件产业的提升，产业链上下游的互动促进了产业链的深度融合，标准的制定契合了当前技术水平，加速了整个产业的良性发展，保障了产品的质量，降低了产品的成本。

【汽车紧固件发展】

加速汽车紧固件产业发展的五大因素，如图1所示。（1）网联化和数字化促进了产业的升级。随着互联网技术的发展，利用互联网技术，建立起人与人、人与机以及机与机的互联，成为当下的一种趋势。以传统紧固件产业为基础上，实现研发、制造、质量、成本等环节互联互通，打通信息孤岛，打造数字化工厂。目前在江苏、浙江等地，正在兴起打造数字化工厂的潮流，数字化工厂连通了所有生产和管理环节，打通数据壁垒，极大提升了生产效率，缩短了产品生产周期。将整个生产过程与成本挂钩，降低了生产成本，从而提高生产企业的竞争力，在全球产业链中赢得一席之地。（2）贸易争端加速了产业链的本地化进程。全球化是当今世界发展的主流，而逆全球化也在局部地区有着自己的市场。2007年欧盟对中国发起的输欧紧固件采取反倾销措施，2018年中美贸易战中美国对一些紧固件产品征收高额的关税，这些都对全球化业务链带来很大风险。紧固件作为工业产品，虽然单件价值较低，但因数量巨大，在多次贸易争端中都被列入加征关税清单。此外，还有一些不可抗因素的影响，比如目前新冠肺炎疫情在全球肆虐，各OEM（Original Equipment Manufacturer）都在积极应对供应链断供的风险，特别是高附加值的进口紧固件，还不具备本土化的条件。国内产业链需要不断提升自身水平，重视研发投入，抓住机遇，迎接挑战。（3）创新设计推动了产业的健康发展。设计作为产品之源，深刻影响着汽车紧固件产业的发展。设计中不仅包括最新的技术，还深刻影响了下游供应商。随着我

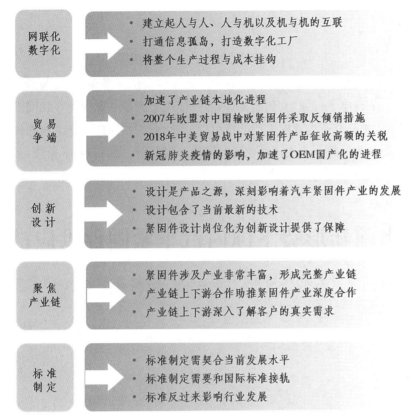

图1　加速汽车紧固件产业发展的五大因素

国汽车工业的高速发展,紧固件设计岗位也在不断规范化,但同时作为小型零部件,国内还没给予足够的重视,一些设计公司缺乏专业的紧固件工程师,导致紧固件产品没有标准化,紧固件种类过多,紧固设计不够恰当,直接影响制造体系、物流体系、质量控制、成本控制、售后服务等,进而导致汽车在行驶中出现各种异响、失效等故障,甚至危及行车安全。同时,作为零部件供应商,也需要不断地进行技术创新,主动为客户提供解决方案,最终形成自己的数据库,供客户进行选择,形成一种良好的创新局面。(4)聚焦产业链上下游。紧固件虽小,但工艺非常丰富,从原材料、拉拔工艺、冷镦、搓丝、热处理、表面处理等,以及模具制造加工,形成了一个完整的产业链,每一个环节都会影响到最终产品的质量和成本。如上一道工艺处理得当,后续工艺就不需要再重复进行,减少的工艺就是降低的成本,降低的成本又会转化为竞争优势。产业链的每一个环节都需要非常了解客户的真实需求,而不是随意地控制每一个工艺点,要识别哪些是关键工艺或参数,把好钢用在刀刃上。(5)标准是质量的保证。标准是行业对某一领域认知的浓缩,进而反过来引导行业的健康发展。制定标准一方面要参考国内生产水平,同时也要与国际标准接轨,过松、过严的标准不利于推动行业的发展,适合的才是最好的。例如某个零件设计采用ISO标准,国内想使用国内牌号替代,但国内该牌号对应标准过于宽泛,结果是该标准牌号不能满足ISO标准,只能使用进口原材料,对于我国的材料行业发展非常不利。

【结论】

号称"工业之米"的紧固件,是我国制造业发展必不可少的零部件,对我国制造业的发展意义非常重大。小到电子行业,大到能源行业,都使用了大量的紧固件。我国汽车行业拥有巨大的保有量,并且具有大规模生产的特点,汽车行业的紧固件对整个紧固件行业都有很大的推进作用。新形势下紧固件产业链的发展,必然依托当前的技术条件,契合当前的全球环境,为汽车行业的产品品质保驾护航!

浅谈当前新形势下国内汽车紧固件供应链

钟云龙*,潘黎,孙勇,刘明华,刘岩

内德史罗夫紧固件(昆山)有限公司

* 通讯作者:yunlong.zhong@nedschroef.com

【摘要】

由于新冠肺炎在全球的继续蔓延,世界经济也受到了巨大的影响,全球经济贸易、实体经济均出现短暂停滞,金融风险也进一步加剧。中国的经济也不可避免地受到影响,国际政治关系也更加复杂。

作为全球和中国经济的支柱产业之一,汽车工业不可避免地受到疫情的冲击。回首过去 10 年,全球汽车销量在 2010—2017 年间持续增长,2018 年略有下滑,而 2019 年则出现明显跌幅。中国汽车销量和全球的汽车销量走势一致,2019 年同比下降 8.2%。目前市场预测,受新冠肺炎疫情的影响,全球汽车市场将出现 15% 的跌幅,回落至 2010 年的水平。第一季度,国内汽车销量同比去年下降 45%,全年预计下降 15% 左右。

受疫情影响,欧美很多车企和零部件供应商暂时关闭,全球的供应链和业务发展面临严峻挑战。此外,由于国际形势的变化,部分欧美企业和日本企业也可能会回归本国或转移到其他东南亚国家。但目前我国一部分紧固件仍然依赖于进口,国产化仍面临一些困难。在新形势下,耐高温紧固件、铝制紧固件、高精密异型紧固件,以及一些国外专利产品都存在一定的供货风险。本文结合内德史罗夫的实践,分析了上述紧固件的现状以及国产化所面临的难点和挑战。当然,挑战的另一面就是机遇。在当前形势下,汽车主机厂和零部件厂将逐步加大关键紧固件的国产化力度,推动国产紧固件产品高端升级。同时,国内企业也存在潜在的并购投资机会。

针对上述问题,本文粗浅地提出了一些相应的建议措施,见图1。国内企业需要整合上下游的技术能力共同提高,包括原材料、热处理、表面处理、设备、模具等厂商。同时,紧固件企业从设备、研发技术、模具设计等方面,逐步提升技术能力,减少对进口件的依赖,降低供应链风险。

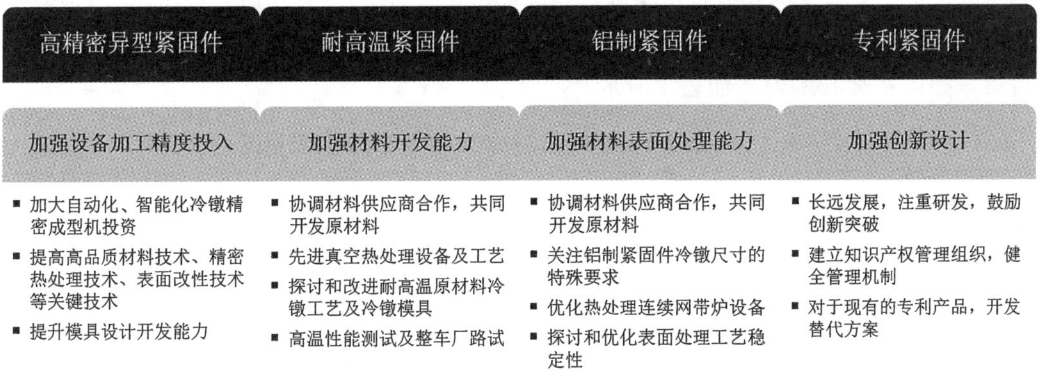

高精密异型紧固件	耐高温紧固件	铝制紧固件	专利紧固件
加强设备加工精度投入	加强材料开发能力	加强材料表面处理能力	加强创新设计
■ 加大自动化、智能化冷镦精密成型机投资 ■ 提高高品质材料技术、精密热处理技术、表面改性技术等关键技术 ■ 提升模具设计开发能力	■ 协调材料供应商合作,共同开发原材料 ■ 先进真空热处理设备及工艺 ■ 探讨和改进耐高温原材料冷镦工艺及冷镦模具 ■ 高温性能测试及整车厂路试	■ 协调材料供应商合作,共同开发原材料 ■ 关注铝制紧固件冷镦尺寸的特殊要求 ■ 优化热处理连续网带炉设备 ■ 探讨和优化表面处理工艺稳定性	■ 长远发展,注重研发,鼓励创新突破 ■ 建立知识产权管理组织,健全管理机制 ■ 对于现有的专利产品,开发替代方案

图1 我国高端紧固件发展方向及建议措施

中国金属学会团标《冷镦和冷挤压用钢》要点解析及紧固件上下游匹配

陆恒昌[1,*],王立新[2],张波[3],左锦中[4],赖建明[1],袁峰[2],董瀚[1]

1 上海大学　2 泛亚汽车技术中心有限公司　3 上海海德信金属制品有限公司　4 中天钢铁集团有限公司

* 通讯作者:luhengchang@shu.edu.cn

【引言】

紧固件是应用量最大的机械基础零件,被誉为"工业之米"。发展紧固件产业技术创新基础设施符合我国的"新基建"政策和发展趋势。目前,我国紧固件年产量居世界第一,如2018年产量超过800万吨,产值达820亿元。然而,我国紧固件行业正处于"中低端产品产能过剩、高端产品供给能力不足"的局面,高端紧固件及关键材料严重依赖进口,制约着我国装备制造业的发展,需要大力发展我国紧固件产业技术创新基础设施,建立上下游匹配的产业技术联盟。团标是紧固件产业技术创新基础设施的重要内容,为此由上海大学牵头联合30多家单位共同制定了《冷镦和冷挤压用钢》团体标准,该团体标准涵盖了紧固件制造的上下游企业,具有广泛的代表性。

【原则及主要技术差异】

本团标以促进紧固件上下游相匹配为目的,以"适用性、先进性、互换性"为原则,结合国内钢铁企业工艺技术和装备水平以及紧固件行业的需求,主要从碳含量范围、冷顶锻性能、脱碳层要求及夹杂物等级等技术指标,参考国内外相关的先进标准而制定。与 GB/T 6478—2015 主要技术差异如表 1 所示。

表 1 本标准与 GB/T 6478—2015 主要技术差异及原因

章条编号	技术性差异	原因
6.1	碳含量范围收窄,组距由 6~8 缩小为 5;对于有淬透性要求的钢种,要求碳含量成品化学成分允许偏差为 0.01%	碳含量波动会造成同工艺不同批次热处理后(球化退火、调质等)性能散差较大,影响冷镦成型的锻压力、模具稳定性和寿命及最终零件强度的稳定性
6.1.4	增加说明:非调质型可供应满足相应紧固件等级的其他牌号或化学成分钢材	根据用户的具体要求,非调质钢的牌号和成分可灵活调整,以符合实际需要及技术发展的需求
6.5	① 冷顶锻性能以 1/3 为最低要求,碳≥0.30% 及含铬、钼等合金元素的钢材,试样可先球化退火;② 增加了 1/5 的选择项	ISO4954—2018 要求最低的冷顶锻试验为 1/3,为了增加可操作性,明确了冷顶锻前可以球化退火的情况,同时考虑下游用户的更高需求,增加了 1/5 选项
6.6	脱碳层要求加严,不允许有完全脱碳层	与 ISO4954—2018 要求一致,符合汽车行业的要求
6.8	明确夹杂物的等级要求,并分成两组	根据下游用户的要求

从原料端看汽车紧固件产业链的发展

孙 华*

南京钢铁股份有限公司

*通讯作者:sunhua@njsteel.com.cn

【引言】

在全球化进程中,欧美形成了产业链上各个环节的统一标准,通过技术标准和严格的供应商准入制度,将汽车厂与不同国家和地区的零部件厂、加工厂、钢厂等紧密联系在一起。日本在此基础上更进一步,除了技术纽带,还加入了利益纽带。以商社介入以及相互参股的方式把产业链紧密联系在一起,保证了产业链的协同发展和共同受益。当前国内原材料制造水平、产业链联动的现状,提出我国紧固件产业链的发展方向。

【我国紧固件产业链现状】

目前国内原材料状况:制造装备先进,具备生产高端汽车紧固件的条件,掌握了大部分制造的关键技术和生产工艺,部分产品已达到国际先进水平,图1为制造冷镦钢关键技术。产业链联动的现状:① 以产品制造流程形成供应链,互相依赖、互相合作(案例1:宝钢—宝日金属—日泰/长华—大众/通用;案例2:南钢—昆山贸盈—舟山7412/上汽标—通用)。② 通过发展共享交流形成技术共享平台,如南钢与上海大学、南京工程学院、通用泛亚、昆山贸盈各技术单元建立产学研用的技术平台,采用EVI研发先期介入模式,为上海通用汽车成功开发了多个项目。

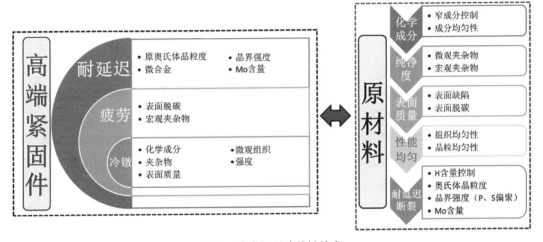

图1 冷镦钢制造关键技术

【结论】

针对当前的疫情影响,中国汽车紧固件的产业链发展应充分借鉴欧美日的成功经验,打造自主的完整的产业链:

各汽车厂、零部件厂、加工厂要敢于使用国内材料。

在现已形成的供应链、技术平台和已有材料基础上,逐步形成各环节的一揽子统一标准(包括产品、交货、包装运输、上货、上下游工艺技术对接等一系列要求),作为发展提升的基础。

各环节协同,为主机厂开发成本更低、替代进口的材料。

寻找全球化机会,参与全球竞争。

考虑资本的纽带作用。

汽车紧固件数字化思考

李大维*

上海汽车集团股份有限公司
* 通讯作者：lidawei@saicmotor.com

【引言】

20世纪90年代以来，互联网的发展经历了一个数字化、智能化的过程，人们日常使用的手机、汽车等也受到了很大的影响。本文分析数字化在汽车上的应用与发展。

【汽车紧固件数字化】

人均GDP越高的国家数字化发展程度越高，目前中国在相关领域的发展速度正在逐渐加快。数字化的发展至少可以为行业带来4%～5%的业务额增加（中科院数据）。数字化的特点包括：(1)数字化营销和销售注重端到端的利用高级分析法来更贴近客户；(2)数字化制造注重利用自动化/机器人物联网、人机界面、穿戴式来提升劳动生产力，利用高级分析法来提高产量和设备综合效率；(3)数字化供应链注重端对端解决方案，连接供应链上的每一方；(4)数字化综合管理注重利用自动化来提升增值活动的占比；(5)数字化采购注重利用高级分析法提高开支效率；(6)数字化研发注重利用高级分析法来提高投资回报率。

近几年来，紧固件行业也在不断推进数字化进程。目前，数字化还处于较初级的阶段，它可以连接汽车生产厂家和设备供应商，如紧固件企业常用的ERP系统。从图1可看出紧固件数字化发展的方向。

图1 紧固件数字化发展方向

【展望】

紧固件行业实现数字化需要具备诸多条件。应用在汽车行业的紧固件在服役后需要对其保持监控,另外,还需要建立全车紧固件系统之间的连接,以便实现完整的数据化,目前已经有汽车能够实现这一功能,其后台可读取试验车的紧固件状态,以此指导紧固件的批量生产,提高紧固件质量控制的效果,这也是数字化理念的应用。紧固件数字化带来的经济效益将变得愈发可观。

新冠疫情下中国汽车紧固件行业的思考

李万江*

东风汽车公司技术中心
* 通讯作者:liwanj@dfmc.com.cn

【引言】

新冠疫情暴发后,我国人民生活、制造业、基础设施建设均无法正常进行,因此,第一季度的国内生产总值出现了负增长,国内经济环境较差。2020年2月份,我国制造业基本处于停滞状态,3月份才开始逐步复工,目前,国外的汽车制造业基本处于停产状态,因此,进、出口汽车的产业链也基本处于停滞状态。紧固件的进、出口业务同样受影响较为严重。

【紧固件产业现状】

当前国内紧固件厂家主要分为国企、民企和合资企业,各类厂家的技术、资金实力参差不齐。合资、外资企业的资金、技术实力较强,民企经营规模较小,现金流可能更容易出现危机,国内紧固件企业的主要角色是"生产工厂",在材料研发方面与合资企业存在差距,在特种紧固件方面仍然存在受制于人的情况。另外,紧固件表面处理技术的供应商也受到了疫情影响,大量中小型企业的经营面临风险。95%以上国内自主汽车品牌的零件都可以实现国产化,受国际产业链影响较小。国内主机厂所需的关键零部件无法从国外进口,国内合资车企面临着停产的风险,加上近几年来国内汽车市场状况不理想,所以部分车企开始裁员、降薪。对于合资品牌来说,大量零件依靠全球产业链的调配,整车产品收国外情况影响较大。目前,合资品牌的主要忧虑在于供应链的变更问题以及国内厂家能否为其提供性能达标、可靠的汽车零部件。

【展望】

合资品牌更多地考虑采用国产紧固件,这对国内紧固件行业是一个好消息,但同时我们应该认识到,国产紧固件在性能、检测、国内企业管理模式等方面仍有很大提升空间。我国紧固件行业的进步需要国内厂家在技术和知识产权方面的积累,也需要国家对市场的调控。

图 1 中国汽车紧固件产业链

新形势下耐热紧固件发展的思考

赵萍丽*,彭其斌

浙江省舟山市 7412 工厂

* 通讯作者:zhaopingli@hj7412.com

【摘要】

本文从紧固件厂的角度,并提出了发展我国耐热紧固件行业的一些建议,分析了我国耐热紧固件的现状及其在新冠疫情蔓延形势下的影响。

【耐热紧固件现状】

国产耐热钢的水平与进口水平仍有差距,主要体现在原材料(成分控制的批次一致性)、材料生产工艺(质量稳定性、一致性、热处理及拉拔工艺)、表面涂覆工艺(涂层选择、涂层稳定性)、服务(响应速度、订单严肃性)、质量控制(钢厂标准统一性)和终端客户认可度等方面。

在我国耐热钢的发展过程中,存在以下几方面的问题:(1)需求分散。材料品种多,需求量较小,对钢厂关注度不高。(2)质量要求高,对钢厂要求更严格。(3)前期投入大,认可周期长。

新冠疫情对耐热钢紧固件的影响:目前国内耐热钢来源如图 1 所示,其中美国进口占比约 20%,日本

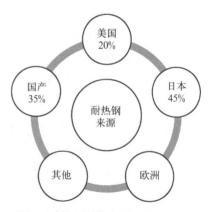

图 1 我国耐热紧固件用钢的来源

进口占比45%，国产占比35%。虽然国产数量占比不小，但主要集中在低端钢种，高端如镍合金钢、铁素体钢、马氏体耐热钢、高温合金钢主要从美、日进口，因此进口产品价值更高。尽管目前国内紧固件厂商已经掌握各类紧固件生产制造技术，但国内大量耐热钢原材料仍采用进口，在疫情期间便产生了供需矛盾。这次疫情使全球需求量收缩，进口材料交货期变长，复产、后序产能受损，运输困难，价格抬升，因此材料国产化是发展国内产业的必要途径。

【对策与建议】

由主机厂、紧固件厂及钢厂三方产学研合作，同步开发，共同推进进程，其中主机厂是关键环节；

统一标准：主机厂图纸指定标准，紧固件厂根据生产情况确定标准，各钢厂执行标准，通过制定标准统一材料牌号命名和生产制造，并对标准进行推广应用和更新；

材料改制、表面皮膜需突破；

材料储存、包装和运输过程需防护；

建立产业联盟，避免价格竞争导致市场混乱；

完善国产耐热材料和产品的市场化和产业化，推动技术、产品、市场的可持续发展。